AGRICULTURAL ENTREPRENEURSHIP

AGRICULTURAL ENTREPRENEURSHIP

Nekesah T. Wafullah

Delve
PUBLISHING

www.delvepublishing.com

Agricultural Entrepreneurship

Nekesah T. Wafullah

Delve Publishing
2010 Winston Park Drive,
2nd Floor
Oakville, ON L6H 5R7
Canada
www.delvepublishing.com
Tel: 001-289-291-7705
001-905-616-2116
Fax: 001-289-291-7601
Email: orders@arclereducation.com

ISBN: 978-1-77361-441-0 (Hardcover)

ABOUT THE AUTHOR

Nekesah T. Wafullah, M.Sc. Agricultural and Applied Economics from the University of Nairobi, Kenya, is skilled agriculture expert with extensive knowledge in agricultural business management services, project management, various forms of fertilizer, their production, sales, marketing aspects, and application regimes; cross-border fertilizer trade policies; youth and women empowerment; and volunteerism. She has experience in agricultural markets in Kenya, Zambia, Malawi, Tanzania, Rwanda, Democratic Republic of Congo (DRC)—Bukavu and Lubumbashi, and Uganda.

TABLE OF CONTENTS

LIST OF FIGURES

LIST OF TABLES

PREFACE

Farmers becoming entrepreneurs is in trend extensively these days. However, what does agricultural entrepreneurship actually mean? An agricultural entrepreneur looks to follow market trends, yield an additional profit; work for the progress of an economy, being an agricultural entrepreneur is more than just having a farm.

Due to lacking entrepreneurship in agriculture, the prospects in agriculture are seized by the outsiders. Considering the increasing unemployment in rural areas and sluggish growth of the agricultural sector, it is essential to take advantage of the opportunities for promoting entrepreneurship in agriculture, which in turn can resolve the existing issues related to agricultural production and profitability. The process of taking advantage of these opportunities of the agriculture business with the potential of each one has recently developed into a study called "agricultural entrepreneurship."

Increased adaptation to the moods of the market is a parameter that always promises more results for the sector. Achieving this aim is best helped by a detailed look through a scientific approach to manage the farmers. To tap into every opportunity for achieving a common goal can be a daunting task and that is precisely what has given room for the development of analyzing the entrepreneurship opportunities into a complete field in itself.

This book will trace how the term agricultural entrepreneurship was coined and its evolution from a theoretical concept to a practical one. There is a plethora of information out there on the development of entrepreneurs but each sector demands a customized approach. The existing development practices can at best give clues on what improvements can be made, but it is the thorough study of individual situational patterns that can give rise to proper strategies that can work in real life.

The range of issues incurred in farming has a common nature but a universal solution cannot be provided for all. Instead, a framework can be developed that can be modified as per the economy's culture. That is precisely what the book will be identifying.

It combines several factors that affect the growth of the agriculture sector to get a clear picture of how the differences in perspective can be made use of.

At the same time, the book will offer very clear insights into the perceptions that can be worked upon for change, which will eventually drive the sector's productivity.

The subject matter of this book starts with establishing a clear picture of the term agricultural entrepreneurship. The education of this approach will contribute to widening the understanding of improving the role of entrepreneurs in farmers, defining the criteria's affecting agricultural development, defining ways to imbibe innovation and curating entrepreneurship qualities. This would be supported by real-life case studies at the end of each chapter to enable the reader for achieving direct results. Next focus would be a list of best practices in agricultural entrepreneurship that have proudly excelled in increasing role of entrepreneurs in farming. This section would also present the existing areas of improvement and challenges included under the various segments aimed at improving the utilization of approaches. Towards the end, a comprehensive detail of the existing challenges would be covered.

SUMMARY

Farmers becoming entrepreneurs is in trend extensively these days Due to lacking entrepreneurship in agriculture, the prospects in agriculture are seized by the outsiders. Achieving this aim is best helped by a detailed look through a scientific approach to manage the farmers. This has recently been developed into a study called "agricultural entrepreneurship." This is a very simple anecdote of the utilization of agricultural entrepreneurship and a complete study has much more to offer.

CHAPTER 1

INTRODUCTION TO AGRICULTURAL ENTREPRENEURSHIP

CONTENTS

Introduction to agriculture entrepreneurship starts with the brief overview about the role of entrepreneurship in the farm sector. The next focus in this chapter is laid upon how in the past, agriculture is eliminated from the entrepreneurial research. Subsequent sections give an overview of the development of connections between entrepreneurship and agriculture. Also, discussed are entrepreneurial opportunities.

1.1. INTRODUCTION

In the past few years, most of the public sectors, prominently the agriculture sector has undergone phenomenal changes.

Due to the recent reforms in the government policies, there is lesser support in the agriculture sector, whereas the rise in retailers in this segment along with the diversified patterns of consumers has opened the need for sophisticated strategies (Figure 1.1).

Figure 1.1. Governmental policies.

Source: https://svgsilh.com/image/1339917.html

However, the farmers have adapted themselves to most of these new changes. This ability of the farmers to apply themselves in entrepreneurial behaviors has led to the increase in several new patterns within the sector.

Expectedly, different disciplines have of late started engaging in the research and development activities for the ongoing changes in the agriculture sector.

The rural sociologists along with agricultural economists studied the impact of these changes on the farming sector and their associated communities.

Further, the entrepreneurship scholars too have largely concentrated their works on the abilities of few farmers to identify and evaluate upcoming opportunities, using farm-based resources to exploit the non-agricultural markets (Figure 1.2).

Figure 1.2. Agriculture markets, Wikipedia (2018).

Source: https://upload.wikimedia.org/wikipedia/commons/5/5a/Milan_Vegetable_Market.jpg

The analysis led to the development of two separate disciplinary approaches, with distinctive origins, paradigms, and methodologies, both of which are concerned with the role of new venture creation in rural economic restructuring, but with little evidence of any linkage between the two approaches.

The objective of these two approaches is to analyze the work of scholars from the enterprise and agriculture research areas, both of which have a common goal – to focus on the entrepreneurship in agriculture, food production, and rural development.

Let us explore the most obvious question in the minds of most of the readers – Why should there be research on entrepreneurship in agriculture and rural development? Considering the entrepreneurship research is performed broadly on the typical conventional industries of emerging technologies that can demonstrate innovation and the prospect of future wealth creation and employment growth, this research on entrepreneurship in agriculture is not very common.

Further, agriculture as a sector is also not known to attract high levels of entrepreneurship. Most of the analysts depend on the support provided to this sector as price support, subsidies, and other market schemes. The crucial factor for their success is not the typical business skills they generally bank upon, but their effectiveness as a political lobby. The agriculture sector is primarily composed of small-scale; family-owned or tenanted farms that have largely survived the transition over generations.

There exists a strong relationship between the agriculture and the entrepreneurship research domains. In spite of the growth of parallel industry within the food and agri-business domain, the small farm agriculture industry has not only withstood the tough competition, but also flourished attributed, largely to the entrepreneurial capacities of farmers and agriculturalists

The objective of this book is to explore entrepreneurship in agriculture, a sector that has been largely overlooked by the entrepreneurship research domain. Entrepreneurship which has majorly contributed elsewhere across domains within diverse agricultural and food-based industries and in distinctive socio-economic and cultural contexts (Figure 1.3).

Figure 1.3. Entrepreneurship in agriculture.

Source: https://www.usda.gov/media/blog/archive/tag/microloans?page=3

1.2. THE ELIMINATION OF AGRICULTURE FROM ENTREPRENEURSHIP RESEARCH

Agriculture has is an industry eluded by entrepreneurship research attention since time immemorial. The factors contributing to the exclusion of agriculture from entrepreneurship research is overwhelming. In the past, the focus of entrepreneurship research has generally been manufacturing, high technology (hi-tech) and the service industries. This focus can be easily justified due to employment growth and formation of new companies.

Most of the government-driven, as well as private business data analysis, have hardly covered the agriculture sector in any of their research projects. Hence, agriculture sector largely remained oblivious of any entrepreneurship research in the domain (Figure 1.4).

Figure 1.4. An adverse effect of drought on farming.

Source: https://pixnio.com/people/male-men/drought-affects-health-and-farming-madagascar

The research done in the UK exhibits most of the government data sources, including the Census of Production, have excluded agriculture. Even the Census of Employment does not cover any data points from horticulture or agriculture domain.

In the USA, the annual State of Small Business Report to the President reports on a business population that comprises 'non-farm sole proprietorships.'

Non-exclusion of agriculture from any enterprise having commonly used data points imposes severe challenges for researchers who would under such circumstances, base some of the specifications pertaining to the agriculture subject with a predominantly substantial set of invalidated assumptions.

The exclusion of agriculture from entrepreneurship may be convenient, but it is theoretically perplexing. Some of the leaders of entrepreneurship theory, such as Richard Cantillon, Jean Baptiste Say and John Stuart Mill, were writing at a time when agriculture was the predominant industry and the examples they used to illustrate their ideas concerning entrepreneurial rents were typically derived from the practices apparent within contemporary agricultural production and distribution (Figure 1.5).

Figure 1.5. Quality check by agriculture entrepreneur.

Source: https://pixnio.com/people/male-men/mexicos-agricultural-programs-to-improve-various-crop-production-for-farmers

Cantillon's (1755) explanation of an entrepreneur as 'An agent who contracts with suppliers at known prices in order to produce goods that could be sold later at uncertain prices,' illustrates perfectly the uncertainty-bearing activities of eighteenth-century peasants who would buy seeds at a known price to grow produce which was later sold for an unknown return.

As a result, the basis of contemporary entrepreneurship theorizing is dependent to a considerable extent upon eighteenth-century agricultural practices.

While it is evident that the entrepreneurship research domain has excluded agriculture sector, the domains of agricultural economics and rural sociology have been equally gradual in realizing the intense developments

This can be explained in part by conventionally held views of agricultural specificity; seen as a 'special case,' conventional farming was hardly seen as a trade let alone an entrepreneurial activity. The periodical of Gasson et al.'s (1988) determining review of farms as family businesses was a turning point in the way that agricultural enterprises were seen by some scholars.

The colleagues of Gasson and she herself argued that the tripartite model of agricultural relations had been removed with the demise of laborers and property owners, leaving a residual farms mass, as much as 90% of which were family headed or worked and tenanted, mostly, by family labor.

According to the Hill's argument (1993), it was risky to assume family farms and small farms as synonyms. It was obvious that the small family farm firms had much in common with non-farming firms. Family farms were featured by a lack of separation between control and management; profit maximization objectives balanced by the requirement for succession and independence; decision made family members; family used as a risk capital and source of labor; and the presence of a strong independence ideology, tradition, and self- sufficiency (Gasson et al., 1988).

During the 1990s, the agro-policy reforms impact focused both on the variety of farm-based activities that could replace diminishing gains from agriculture and on the farm household as a supplier of ideas, labor, and resources for new firms and activities (Shucksmith and Herrmann, 2002; Fuller, 1990).

While most research has considered the activities, motivations, and performance of fresh farm- based industrial enterprises, the move towards 'post-productivism' and the advent of multifunctional agriculture (Evans et al., 2002; Potter and Tilzey, 2005); the farm-based business venturing analysis has been performed mainly without reference to the entrepreneurship research realm.

1.3. THE SEPARATION OF AGRICULTURE AND INDUSTRY

The split between industry and agriculture is much broader than is evident within the entrepreneurship realm. Within the fields of economics, sociology, geography, and history, agricultural manufacturing and the rural environment in which it occurs are considered distinct subjects for teaching and research.

Though the entrepreneurship literature has focused on the industrial firm to the agriculture exclusion, other disciplines have deeply considered the agricultural sector, but as a distinct and separate field of inquiry from industrial growth. The partition of agriculture from industry is not simply an outcome of the research of scholars; the partition prevails within the political, economic, and social institutions that surround the sector.

Jointly, these institutions reflect a split between other economic activities and agriculture that strengthens a view of agricultural uniqueness and 'specificity.' The immense requirement for food security as an issue of global and national importance, combined with the ability of strong interest groups to activate political support, safeguard that agriculture has been seen often as a 'special case' (Figure 1.6).

Figure 1.6. Food security.

Source: https://www.flickr.com/photos/codnewsroom/41582048164

At the economic level, this is manifested in the support that has been given to institutions responsible for administering aid to, and overseeing the efficient functioning of all agriculture and agriculture-related enterprises.

At the social level, issues relating to land tenure also separate agriculture. The land is isolated from other forms of capital in as much as it is in fixed supply, is subject to climatic and topographical constraints and, mainly, is connected intricately to a relationship structure of stewardship and ownership.

Though there is an argument about the beginning of the industry and agriculture separation, apparently it is not a recent phenomenon. Kautsky's (1988) The Agrarian Question analysis, first printed in 1899, explained the separation between industry and agriculture as emerging in the early medieval period.

These societies were stable, self-sufficient, and calm comprising of peasant artisans. The advent of small industries began a process of commercialization of agricultural products and increasing specialization, which led to the separation of the two forms of production.

While a comprehensive analysis of the separation origin is beyond the reach of this introduction, The Agrarian Question is specifically important for different reasons. Kautsky's historical analysis of the peasant artisan offers striking insights into many of the modern issues discussed within the chapters of this book. Kautsky's explanation of the series of activities carried out by medieval peasants presents us with a prior example of what is presently termed as farm diversification, and his analysis of peasant proletarianization process, the amalgamation of small-scale agricultural growth with off-farm employment, offers a theoretical framework for the analysis of farm-based pluriactivity (Bryden et al., 1992).

Though it is alluring to view industry and agriculture as distinct socio-economic spheres, this negates the relationship complexity between the two production forms and the historical links between urban and rural societies. While modern analysts often look agriculture as having only a small status as compared to industry, in the past agriculture was seen as vital, not only in terms of productivity and employment but also in its role in the growth of industry and urbanization.

As Adam Smith (1776) discussed in The Wealth of Nations, 'The great business of every educated society is that carried on between the population of the town and the country. The country provides the town with the means of survival, and the resource for manufacture. The town reverts by sending manufactured produce to the country's population.

Extra agricultural productivity brought about the eighteenth and nineteenth centuries industrialization, along with the increase in urban residency. As urban residence became popular, popular concepts of the sector in the form of price support, production, and other market regulatory procedures.

The postwar beginning towards agricultural production, initially introduced as a means of safeguarding food supply security, is deeply ingrained within the sector. Agriculture is also marked by the prevalence of specific political institutions.

A common characteristic of both developing and developed economies is the prevalence of focused legislative rural life became romanticized increasingly, a view obvious in the use of Gemeinschaft (community), Tönnies' ([1887] 1955) terms and Gesellschaft (society) as polarized descriptors of urban and rural (Harper, 1989). While the past views regard industry and agriculture as distinctive and separate spheres, what of the modern view. The advent of large-scale agri-business, vertical integration within the food chain and the implementation of industrial management style in agriculture have not only upgraded traditional practices, they have end formerly rigorous class positions and modified rural communities. The impact of these changes has not been spread evenly throughout the sector and current agriculture is as much featured by its diversity as it is by its intricacy (Figure 1.7).

Figure 1.7. Care farming.

Source: https://upload.wikimedia.org/wikipedia/commons/a/aa/Students_learn_how_to_weed_in_the_specially_designed_wheelchair_accessible_garden_beds._%2810727065023%29.jpg

Largely, the foundation of agricultural 'specificity' has long since battered. Indeed, 25 years ago rural sociologists argued that 'the answer to the question, what is specific to agriculture under capitalism, is nothing.' It is not that agriculture has developed parallel to industry, but that links in complex chains connect production of specific plants and animals to equally specific manufacturing processes' (Friedmann, 1986: 44).

Given the deep changes that took place within the agriculture sector in the prime period, the question arises as to whether the separation of industry and agriculture still prevails anywhere beyond the sphere of academic subject domains.

1.4. EVOLVING INTERACTIONS BETWEEN ENTREPRENEURSHIP AND AGRICULTURE

Clearly, entrepreneurship has always been an important feature of the agricultural sector, a fact often hidden by both the prevalence of complex market regulatory mechanisms and the astonishing presence of the belief that agriculture is a unique case to be emphasized separately from other forms of economic activity.

Though the development of scholarly specialization has meant that there has been little cross-fertilization to date between the domains of entrepreneurship and those of agricultural economics and rural sociology, there are strong arguments that this should change. The deep changes experienced by the agriculture sector as a result of policy liberalization and demand-side changes and, in particular, the acknowledgment of farmers' 'unequal tendency to adapt and adjust' (Shucksmith and Herrmann, 2002: 37) has drawn focus to the requirement to better understand changes in individual nature and farm strategies.

In spite of critiques of the shift towards a 'post- productivism' research orientation (Evans et al., 2002), the requirement to understand the essential procedures that drive farm household planning is supreme.

In this regard, the entrepreneurship domain provides significant research experience and deep understanding that can be applied profitably to the farming sector analysis. At the same time, the research area of entrepreneurship has much to learn from the agriculture sector and from the agricultural research area. The small family farm firm is similar to the non-farm firm and these similarities are likely to grow over time as stress expands on farm incomes.

As such the sector provide entrepreneurship researchers a distinctive opportunity to assess issues that are essential research themes within the realm, like the family–business nexus, the organization of portfolio entrepreneurship, the exploitation and identification of opportunities, and the role of the family in business growth and performance, and business ownership over multiple generations.

The inspiration for this edited research collection on entrepreneurship in rural and agriculture development is to better understand the benefits that may exist between the two subject realms and to explore the gains of further studies. In this regard, this book must be considered as a means of evaluating the water for future collaborations.

The aim is to both discover the contribution that the entrepreneurship research realm may offer to the arguments covering rural and agricultural development and to consider whether and in what ways a consideration of the issues covering the agriculture sector can light up some of the less well understood aspects of entrepreneurship practice and theory.

We now set the background for the following book chapters by providing some opening thoughts on the cable connections that we see between the two subject realms, in particular, the emphasis on entrepreneurial resources and opportunities that may offer an alternative theoretical lens via which to view some of the important debates within modern rural and agriculture studies.

1.5. THE ATTENTION ON ENTREPRENEURIAL OPPORTUNITIES AND RESOURCES

Some as consisting of three unique but overlapping views (Alsos, 2007) have explained the entrepreneurship domain. The innovation-based view derives from the Schumpeter work, who see entrepreneurship as the building of an imbalance in the market based on a new blend of resources, in the form of new processes, new products, new organizational solutions and new markets (Johannisson and Landström, 1999).

From the point of view of business formation entrepreneurship is the process of building new business organizations, from entrepreneurial intention to the establishment and development of new organizations.

This method of organizational creation is termed as the entrepreneurial process (Bygrave and Hofer, 1991) or organizational emergence (Gartner, 1993). The opportunity- based view places the opportunity pursuit at the center of entrepreneurship (Shane and Venkataraman, 2000) (Figure 1.8).

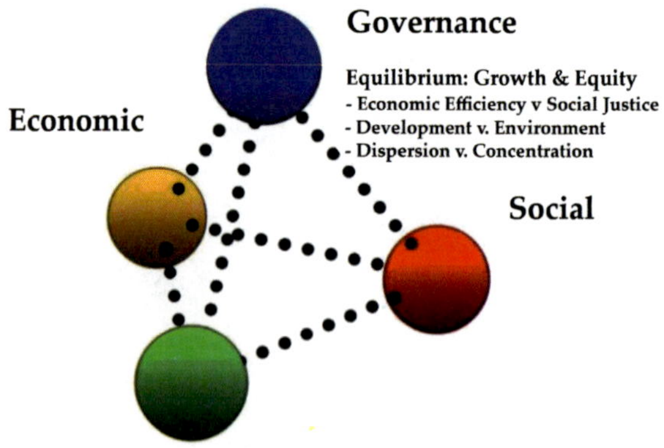

Figure 1.8. Entrepreneurial progress.

Source: https://en.wikipedia.org/wiki/Metropolitan_Reticular_Matrix_Planning#/media/File:The_Tripod_of_Reticular_Matrix_planning.jpg

Studies usually take as a beginning point Kirzner's (1973; 1997) view of the 'alert entrepreneur,' who recognize business opportunities as imperfections in the market and manage resources to exploit these opportunities, thereby restoring market equilibrium (Landström, 1999).

From this viewpoint, entrepreneurship is termed as the invention and exploitation of business opportunities (Shane and Venkataraman, 2000).

The opportunity- based view of entrepreneurship looks specifically important within the perspective of the agriculture sector, implying an entrepreneurship view being concerned with the formation of new business activities via the detection and exploitation of opportunities of which the management of resources constitutes an important part (Alsos, 2007).

The emphasis on new entrepreneurial activities implies that the outcome of the business process may be a new business activity or a new firm within an existing firm.

While farms are, of course, commercial zones in their own right, the emphasis of policy and research attention has been on the potential of farmers to produce additional opportunities, arranged either as new business ventures or as part of the prevailing business entity (Bryden et al., 1992).

The concept of pluriactivity, explaining farmers' involvement in revenue-generating activities in addition to conventional agricultural production, is similar to the concept of portfolio business, the ownership of different businesses, studied within the business domain (Carter, 1998; Alsos and Carter, 2006). Portfolio owners have been identified as an essential type of owner, more attentive to business opportunities and more able of efficient searching of opportunity (Ucbasaran et al., 2001; Westhead et al., 2003) (Figure 1.9).

Figure 1.9. Revenue generating activities of farmers.

Source: https://blog.ipleaders.in/agriculture-income-types-taxability/

They are expected to place more focus on new creation, have more goal-oriented and individualistic personal attitudes and are considered more capitalistic than other types of a business industrialist (Westhead and Wright, 1998). One of the main confusions that research relating to business opportunities has dealt with is why, when, and how some individuals produce opportunities while others do not (Shane and Venkataraman, 2000).

There seems to be harmony that some people are more able to recognize particular opportunities than others, and the discovery of that opportunity demands the control of the necessary information as well the innovative ability to access this information (Gaglio and Katz, 2001; Corbett, 2007).

Opportunities differ largely in characteristics and difficulty, and so do the method via which they are recognized. There appear to be at least three extents of such variations. Firstly, opportunities may be seen as either a Kerznerian or a Schumpeterian type (Shane, 2003). For Schumpeter (1962), industrialist used new information, an outcome of research or the new knowledge generation, to regroup resources into more precious forms.

Distinctly, as per the argument of Kirzner (1997) opportunities prevail due to variations in access to prevailing information. Market Knowledge, how to serve markets and knowledge about customer problems are mainly influential factors influencing an individual's attentiveness to opportunities. Secondly, opportunities may occur serendipitously or via planned searching (Gaglio and Katz, 2001).

The view that industrialist recognizes opportunities as an outcome of better information processing ability, scanning behavior or search methods put investment in information in the mid of opportunity search (Fiet, 1996). Thirdly, there is also disagreement regarding whether opportunities exist objectively and require to be discovered or whether opportunities are created subjectively and designed by the industrialist (Gartner et al., 2003; Sarasvathy, 2008).

The recognition of opportunities has been recently seen as a method, which incorporates elements of both 'subjective' creation (McMullen and Shepherd, 2006), and 'objective' recognition. While opportunity recognition and creation are the initial points of the industrial process, utilization of the opportunity is the next stage. Successful opportunity utilization normally relies upon the individual's potential to mobilize or access the required resources to bring the idea to execution.

Industrial resources are normally defined as the immaterial and material assets owned by the firm (Penrose, 1959). Firm resources are usually seen as including all of the assets, organizational processes, capabilities, firm characteristics, knowledge, and information managed by the firm, which makes the firm able to produce and implement strategies that give it an aggressive advantage (Alvarez and Barney, 2000).

Barney (1991) divided resources in three categories: physical capital resources, including premises, geographic location, access to raw material, equipment, and physical technology; human capital resources, including judgment, intelligence, training, experience, insights of the individuals and relationships; and organizational capital resources, including the firm's structures and systems, as well as familiar relations among groups within the firm and between the firm and the environment. Alvarez and Barney (2000) added industrial capabilities to this list of resource categories.

Industrial assets, including Knowledge, learning, and creativity are subtle but can be seen as the ability to identify which resources are or will be important, adapt, and expand such resources, bind resources in useful bundles and set them for the tasks where they will be used most efficiently. As Penrose (1959) suggested, the prevalence of precious resources alone is not enough to create a competitive advantage – how they are utilized is also important.

In the case of portfolio industrialist, or pluriactive farmers, their prevailing businesses may act as a foundation for the development of new business activities (Carter, 1998; Alsos, 2007). Portfolio businesspersons can begin to expand new business activities on the resources owned within their existing firms, which may diminish the problem linked to the building of an initial resource base. An existing firm offers access to resources like employees, organizational routines, customers, and suppliers, as well as physical resources like equipment and buildings (Alsos and Carter, 2006).

This may give portfolio owners a significant advantage over beginner entrepreneurs who have yet to acquire the required resources to begin business. Starting a fresh business from a prevailing business base may diminish both the riskiness of the new venture and by distributing resources between the business activities, the expense of a new startup.

Studies that have emphasized on resource distribution between business activities within businesspersons' portfolio have usually assumed that gains would flow from this type of resource transfer.

Inherent in this approach is the perception that resources are viewed as assets able of contributing to competitive advantage. Currently, it has been identified that some organizational resources may be costly, immobile, and constitutes a liability for the new firm (Mosakowski, 2002; Alsos, 2007).

Immobile and unsuitable resources are not only evident within the organizational and physical capital resources; they may also be prevalent in human capital resources. Skilled entrepreneurs initiating new business activities may also accustom liabilities resulting from their prevailing firms.

Starr and Bygrave (1992), for example, marked the context dependency of entrepreneurial startup information. In addition, experience liabilities may stop from stagnated networks, a reluctance to learn, fixed capital and other resources and diminished time availability.

Reuber and Fischer (1999: 31) argued that some resources could be both liabilities and assets simultaneously, 'For example, a long stay in a particular industry may permit an individual to perform more effectively or efficiently while starting a new business in that industry, but also may restrict that individual from looking new alternatives or opportunities.'

The entrepreneurial actions analysis from a viewpoint that marks opportunity identification and exploitation, and the assets available to the entrepreneur has been a main fruitful line of research within the industrial sectors. The partition of agriculture from other forms of the industry has safeguarded that the farm businesses analysis has taken place in a different academic sphere, and could not come to the notice of scholars familiar with paradigms, theories, and approaches that have been set within the realm of entrepreneurship research.

We believe that the entrepreneurship research realm has a very precise contribution to make to the understanding of the present changes that are occurring within rural areas and agriculture, and the chapters contained within this book explain some of the ways in which entrepreneurship methodologies and paradigms can be applied to the analysis of these deep changes.

1.6. ENTREPRENEURSHIP IN THE FARMING SECTOR

There are three main parts within this book. Primarily, the first part focuses on the entrepreneurship. This particular section primarily focuses upon the organizations and individual entrepreneurs, researching upon collection entrepreneurship within the agriculture sector (farm-based pluriactivity), the antecedents and concerns of multiple business ownership, its resource necessities, and the effect and belongings of the farm household.

Some of the chapters of this book offer evidence about the heterogeneity of entrepreneurial actions in the agricultural sector. Agricultural based ventures vary in the degree to which they are linked to other farm activities, categories of products or services distributed, in addition to their financial impact. There are several reasons for that and it may vary, while on the other hand, be linked to the entrepreneurs on their own such as their motivation, their competence or abilities.

Alternatively, the resource fruitfulness and the categories of resources obtainable at the farm too play a significant role. These particular literature discourses pull factors in comparison to push factors such as motivation for new business start-ups. Even though push factors are frequently seen to be linked to need entrepreneurship, these are created in developed and developing countries both, though the level of necessity may fluctuate.

Upcoming chapters very clearly explain a mixture of pull and push factors, which instigate motivation of the agriculture, based entrepreneurs. There may be chances that entrepreneurs may pick entrepreneurship to diverge from a displeasing status quo. However, in order to be successful in these tasks, they should be able to recognize chances and send resources to use these chances. The capability to do so is reliant on the present circumstances of the farmers in question.

Next focus demonstrates the ways in which family far, as an idea may put limitations together with it may offer chances to the growth of farm-based ventures. In the times, when families are facing transition periods, the option to locate new chances and to act according to them may be bigger in more stable times. Easy access to resources plays an important role in all-new business venturing.

There are significant important resources that are offered by the farm and those can be utilized in the growth of new ventures. The farmers, who are able to redirect present resources, can use the farm as a seedbed for the new ventures in its initial times. Therefore, resource-rich farmers have a benefit.

However, some of the parts show that profit is not attained only based on resources, but also by a capability to understand the possible exclusivity and new uses of current resources. For this, some extent of entrepreneurial approaches and alertness is desirable. Next part emphasizes on change within forestry, an industry involving of several small and a few larger enterprises.

In addition, it is contended that the entrepreneurial perspective is bigger among the larger firms, as they have greater resources. Entrepreneurship in the farming sector is considered as a political topic from sometime and it is seen as a method to change the agriculture sector together with to bring new economic development to rural areas. In addition, there is a political debate linked to the requirement to improve the entrepreneurial activities of farmers, a group which conventionally has not been focused towards the market.

Despite the fact that entrepreneurship has a comprehensive appeal among farmers, they discovered that entrepreneurial abilities are not always measured appropriate to them, since many farmers operate in settings where they do not have the potential to involve in activities that necessitate such skills. Again, heterogeneity should be recognized. Changes in activities and business settings may reason for the substantial difference concerning the extent and ways in which entrepreneurship discourse are accepted, internalized, and reconstructed.

1.7. CONCLUSIONS

Concluding from the different lessons from this book was not an easy task; however, it provides clear subject all throughout the document. One of the instant observations worries the heterogeneity that is present among farm-based ventures. Variety can be observed among the entrepreneurs, who vary with respect to their inspirations to begin new ventures, the resources they regulate, the method they learn, their ambitions and aims, the categories of products and services they grow and the setting in which they function. Regardless of background differences, their geographic, and market putting, the development of farm companies needs considerable entrepreneurial abilities.

These abilities can be seen in the talents of various farmers to encourage their idea-generative capability and to test themselves in discovering the prospective exclusivity of their resource base. Even though some individuals may look to have strong inborn abilities, the popular obtain entrepreneurial abilities through practice. Closer relations with markets and customers, either face-to-face at farmers' markets and through collaboration with other actors in the value chain, or virtually through the utilization of ICT, offers opportunities for farmers to improve an entrepreneurial alignment in addition to a marketing orientation.

Notwithstanding the common encouragements for farmers to develop more entrepreneurial, a comprehensive understanding of their truthful capabilities in this respect also necessitates a gratitude of the restraints that farmers may face. As the following chapters demonstrate, the family farm puts limitations in addition to providing opportunities in the growth of farm-based ventures. It seems that new opportunities may be additionally frequent in the identification and are exploited throughout makeshift periods in the family life cycle, rather than during periods of comparative constancy.

Even though it is identified that farm-based entrepreneurship poses several issues for practitioners, it similarly necessitates an urbane policy framework to guarantee the suitable support at local, regional, and international levels. Policies to support farm-based entrepreneurship are chiefly interesting given the heterogeneity both of the sector and of individual experiences. While governments have established significant experience in motivating and supporting entrepreneurship within small and medium-sized enterprises (SMEs) in non-farm sectors, this experience has not always been transferred into the farming sectors.

One example of this lies in the role that policies may play in reducing the risk associated with entrepreneurial venturing. Within the non-farm sectors, the provision of start-up aid in the form of seed capital and business advice and support has been widely adopted, largely as a result of active engagement and dialogue between policymakers and SME owner-managers and their representative organizations, but the support for farm-based ventures appears more piecemeal – largely as a consequence of the multiple and sometimes competing objectives apparent within agricultural and rural policy.

A recurrent motif within the following chapters is the recognition that farm-based entrepreneurial ventures offer benefits to the wider rural community as well as the individual farm family. Entrepreneurial ventures within the agriculture and forestry sectors often arise because of community actions, suggesting that an important policy dimension lies in supporting local entrepreneurial capacity building.

The following chapters, which focus on different aspects of the entrepreneurial process within agriculture and rural areas, draw on data from a variety of different contexts and using a variety of different methods. While these studies provide many insights into farm-based business venturing, many questions remain unanswered. We view this collection, as the first step in what we hope will be a long and fruitful collaboration between the domains of agriculture and entrepreneurship.

REFERENCES

1. Carter, S., Welter, F., Ljunggren, E., & Alsos, G., (2011). *Introduction: Researching Entrepreneurship in Agriculture and Rural Development*. [eBook] Available at: https://www.researchgate.net/profile/Gry_Alsos/publication/235966675_Introduction_Researching_entrepreneurship_in_agriculture_and_rural_development/links/02e7e51aa47c2339e7000000/Introduction-Researching-entrepreneurship-in-agriculture-and-rural-development.pdf [Accessed 30 Aug. 2018].

CHAPTER 2

AGRICULTURAL ENTREPRENEURSHIP: RESPONSES AND QUALITIES

CONTENTS

This chapter discusses about the economic situations and difficulties faced by agripreneurs. It likewise talks about generic entrepreneurship, entrepreneurial networks, systems, and innovation using a list of opportunities for entrepreneurship, bolstered by agribusiness related cases. The purpose of this chapter is to provide a better understanding of the concept and practice of entrepreneurship. With this understanding, it is trusted that extension workers will be better able to help farmers develop an entrepreneurial spirit and skill set: skills and spirit of an entrepreneur.

2.1. INTRODUCTION

There are more than three billion individuals working around (World Bank, 2012). Of the three billion individuals with occupations, cultivating, and family unit smaller scale organizations give around 1.5 billion business openings all around with half of the work in creating nations.

Occupations are particularly vital for expanding quantities of underemployed youth in creating nations where there will not be sufficient employment to utilize the expanding number of jobless. In these nations, the biggest gathering of those individuals is youth matured 15–24 (40%) (Kapsos, 2013).

A lopsided number of those individuals are youngsters, who will go after ever less occupations later on, as employment creation slacks populace development (Jones, 2015). Accessible employments will require training and abilities that needy individuals do not have.

The few occupations open to individuals without an instruction or applicable aptitudes do not give bearable wages. Business enterprise is a reasonable technique for upward versatility, as a 1% expansion in entrepreneurial exercises diminishes the destitution rate by 2% (Singh, 2014).

Likewise, entrepreneurship is a method for recognizing value-added activities that expand proficiency and work openings. Business visionaries with fruitful organizations are independently employed. Subsequently, business enterprise diminishes joblessness (Gorgievski et al., 2011). Entrepreneurship is especially imperative to agribusiness for a few reasons (Milestad et al., 2011).

A few difficulties to agri-entrepreneurial accomplishment in provincial territories incorporate feeble instruction frameworks, constrained access to mail, and rivalry from extensive homesteads with innovation that expands their profitability. Stenholm and Hytti (2014) separate the business visionary rancher from the maker agriculturist. They recommend the businessperson ranchers set up their personality by tending to challenges in institutional standards and going about as change operators (Figure 2.1).

Figure 2.1. Urban agriculture.

Source: https://en.wikipedia.org/wiki/Urban_agriculture

An expert in agriculture focuses on following standard conduct. Since there are such a significant number of meanings of agri-business visionaries, in this part, the agri-business person will be dealt with as a creative change operator who discovers chances to utilize land and its assets for specific reasons, esteem included food and Agri-based business organizations.

The individual expects all dangers and advantages got from the business. Agri-business visionaries claim and deal with their business with the expectation to make benefit. Agri-enterprise, largely, is a blend of farming and business (Bairwa et al., 2014).

This blend of farming and business cultivates agri-business people who advance, recognize advertises, and fulfill needs by creating diverse methods for getting things done. Agri-business people take part in an assortment of exercises in agribusiness and related areas. Agriculturist business people see their homesteads as a business. They see their ranches as methods for winning benefits. They are energetic about their homestead business and will go for broke to make their ranches beneficial and their organizations develop.

2.2. AN APPROACH TOWARDS AGRICULTURAL ENTREPRENEURSHIP

In the worldwide setting, food security and nutrition is still turned into a critical issue. The significance of food nutrition depends on the Food and Agriculture Organization of the United Nations (FAO), that during the time 2010–2012, 870 million of 7 billion of total populace 12% of them experiencing the absence of sustenance account (interminable undernourishment or craving).

Then again, UNICEF (2013) additionally revealed that meanwhile, in the matter of sustenance, in 2011 around 101 million youngsters matured under five years of age (little children) in various parts of the world encountered an absence of weight (underweight), 165 million had a few impediments development or diminutiveness (hindering), and 52 million be enduring shrinkage (squandering) on the grounds that a dietary insufficiency (undernutrition) (Figure 2.2).

Figure 2.2. Food nutrition.

Source: https://www.mtsac.edu/nutrition/images/146807105_Food-Groups.jpg

In view of the truth of food issue, World food day on 16 October this year get the topic "manageable food frameworks for food security and nutrition." This topic gives accentuation how vital the food framework maintainability to help food security and nutrition.

It is additionally the case in Indonesia, an endeavor to exemplify the accessibility of reasonable food, to help food security and nutrition quits fooling around consideration. Truly, as of recently, Indonesia reliant on food imports is still high. In light of the information discharged CNN-Indonesia, in the most recent five (years 2009–2014), Indonesia's imports increment volume as well as the estimation of offers.

Likewise, its kind, imports Indonesia isn't just rice yet, in addition, an extensive variety of other foodstuff, some of them are wheat, sugar, soybean, corn, drain, organic products, vegetables, dairy animals, hamburger, wheat flour, salt, potatoes, signed oil, and bean stew.

The advancement of the aggregate volume of imports to those 15 sorts of the products are likewise kept on expanding from 12,366 million tons in the year 2009 expanded to 19,766 million tons in 2013 (or expanded by 60,03 %), and in the period from January to June adding up to 10,9 million ton and the esteem sum US$ 4,9 million.

Likewise, estimation of import additionally kept on expanding from the US $ 5.2 million years 2009 to US$ 9.54 million. Thus, when Indonesia did not can decrease reliance on imports will lose the chance to finance the national economy.

The reliance would increment if not ready to give arrive adequately. As per a forecast in 2017 Indonesia would endure food emergency if the development of populace high (1.3 to 1.5) not taken after by the advancement of land. The reliance will be greater if not skilled land. As per a forecast in 2017, Indonesia would endure food emergency if the development of populace high (1.3 to 1.5) not adjust with the improvement of the land.

To enhance the accessibility of food to go independent level there are numerous exertions done by the administration begin from the extension arrive, water system advancement, manure sponsorship, and bundles in innovation. Yet, utilization of these methodologies more underlines on the strategy development.

Then again, approach including potential inward agriculturists, for example, inspiration, imagination, fearlessness, bravery in the hazard, the individual capacity, which by Priyanto (2006) called as business enterprise never figured. Since ranchers, is likewise agents must be required to having the character as a businessperson. In light of this reason, so in the universal meeting this, essayist pushing a thought the expanded generation of sustenance to go self-sufficient food through agriculturist's entrepreneurship.

2.2.1. A Perspective of Entrepreneurship as a New Approach to Increasing Food Production

It is necessary to understand that agriculturists are simply the primary performers to accomplish sustenance adequacy. Notwithstanding, when taking a gander at chronicled increment in food creation that has been portrayed, all endeavors made by the administration in either towards independence in rice and sustenance independence more accentuation on the parts of development alone (on cultivate).

As it were, endeavors to expand nourishment generation are sought after through an approach that exclusive accentuates the part of development. This appears in an assortment of bundles offered development innovation, in particular, "Panca Usaha, Sapta Usaha, Insus, Inmum, SUPRA INSUS, and others. In usage, agriculturists are inspired, driven, coordinated occasion encouraged to actualize all innovation bundle offered by the government.

In such conditions, more farmers are set as a question of government arrangement and agriculturists must consent. Rather, a superior approach puts agriculturists as subjects with either inner potential free mentality, lively, set out to attempt, incredible want, have a requirement for accomplishment, imaginative, striking danger, have the learning and aptitudes by Priyanto (2006) alluded to as the entrepreneurial soul has never been connected.

The significance of entrepreneurial agriculturists as another approach in enhancing cultivating execution, among others, depends on the entrepreneurial qualities.

According to Machfoed (2004), the enterprise is the trendsetter who can change openings into new thoughts, autonomous mastermind, and think about disappointment as study materials, inventive, and capable in getting ready, overseeing, and estimating hazard.

Concurring Priyanto (2008) a man who has the entrepreneurial soul will demonstrate the state of mind of a free, lively, set out to attempt, extraordinary want, have a requirement for accomplishment, imaginative, striking danger, have the learning and abilities related with cultivating being run.

2.3. RESPONSES TOWARDS AGRICULTURAL ENTREPRENEURSHIP: CAPTURING VALUE WITHIN THE VALUE CHAIN

The greater part of the cash in cultivating is made after agriculturists offer their item. Entrepreneurial farmers effectively search for approaches to catch the additional incentive inside the esteem chain. This will build benefits.

Farmer people understand that capturing worth requires delivering for purchasers and final buyers. Be that as it may, simply creating and offering is inadequate. This requires more noteworthy comprehension and information of significant worth chains and their diverse components. It requires an arrangement for taking an interest additionally down the esteem chain.

By what method would farmers be able to catch more esteem? The most widely recognized route is by offering their general crisp creates specifically for shoppers. Another route is to offer high-esteem items, for example, natural items.

Rather than delivering items that are undifferentiated, agriculturists create wares that are separated, for example, natural organic product, an exceptional kind of meat or other particular items for a constrained "specialty" advertise. A specialty market can likewise be made when delivery is created at a time when a couple of different ranchers can contend. Another method for adding esteem is to go into generation and displaying contracts.

Contracts are progressively being utilized amongst farmers and info providers, agriculturists, and processors and amongst agriculturists and different purchasers for the generation and supply of crisp or handled rural create. Catching an incentive inside the esteem chain requires delivering for purchasers and last customer and this requires learning and comprehension of the chain and its bottlenecks and openings.

2.3.1. Different Types of Contracts

- Market-determination contracts: the agriculturist consents to create to a concurred quality.
- Production administration contracts: the purchaser takes an interest in underway administration through assessing creation forms and determining input utilization.
- Resource giving contracts: the purchaser super tight clamps creation and supplies key information sources. The purchaser regularly possesses the item and agriculturists are paid by volume.

Another approach to catch an incentive along the esteem chain is by incorporating with or becoming tied up with the value adding undertakings along the esteem chain. There are various ways – separately and overall – to do this: Pre-creation ventures

Pre-production enterprises: Start providing or are tied up with organizations that supply inputs and different assets, for example, seed, compost, pesticides, devices, and gear.

Post-harvest handling of products: Start giving or get tied up with organizations that give post-collect activities and procedures that increase the value of items, including cleaning, cooling, bundling naming, bundling, transport, and promoting.

Post-harvest processing of products: Start preparing or become tied up with organizations that process agrarian items into more refined items including:

- Milling maize and different grains;
- Cooking, restoring or drying meat;
- Drying foods grown from the ground;
- Mixing wares, for example, nuts, and raisins;
- Create handcrafts with products, for example, grasses, and blooms.

Each of these represents an esteem including plausibility in which entrepreneurial agriculturists can involve to capture incentive inside the esteem chain. On the off chance that they can arrange the essential accounts, they can set up a business to give at least one of these capacities. In any case, for some new farmer business people, venturing into these regions may not yet be conceivable.

Overall, they will perceive and assess these openings and utilize associations and contacts to discover approaches to make linkages so they can progressively catch an offer of the esteem being including along the esteem chain.

Expansion specialists can assume a noteworthy part in helping farmers recognize, explore, and assess openings in the esteem chain. They can likewise help encourage access into or linkages with these open doors in a deliberate and practical way.

2.3.2. Enhancement of Efficiency in Production

Agriculturist business people ceaselessly look for approaches to enhance proficiency, cut expenses and upsurge in profitability. It could be by decreasing the level of cash put that is put into resources such as equipment or searching for inputs that cost less, however they should give the same or better creation results.

It could be by changing the creation framework to one that is less expensive. This may mean surrendering the manner in which one typically does things – regardless of whether it has been done that way for quite a while. There is an exchange off while changing generation frameworks.

Farmers need to adjust the objective of expanding profitability with the underlying expenses of the speculations required. This requires watchful examination and incorporates thinking about such things as acquiring less information sources and materials, utilizing more on-cultivate assets, (for example, work, and waste items) and improving utilization of gear and apparatus. Another choice is to share chances and lessen proprietorship costs.

An illustration is sharing apparatus possession to diminish proprietorship expenses, and advantage from better mechanical innovation. Shared possession agreements function admirably among agriculturists who have comparable measured homesteads and deliver comparative items.

Under such understandings agriculturists consent to pay their own particular running expenses and offer repair and support costs in the extent to their territory utilize (paying little mind to where the machine separates). Diminished proprietorship costs coming about because of sharing is normally bigger than any yield/wage misfortunes caused by delays in mechanical activities caused by sharing.

2.3.3. Land Management and its Sustainability

Whatever way agriculturists and expansion specialists take after, an inquiry that remaining parts is whether the way will prompt increasingly or to less reasonable land utilize. Solid connections with the national, local or global markets, joined with the craving to build cultivate pay and benefit, may urge farmers to extend the territory under creation or change generation frameworks without finding a way to keep up the profitability of the land.

Notwithstanding, if advanced astutely, expanded salaries could be the best certification for feasible land utilize. Making benefits empowers ranchers to re-put benefits in soil fruitfulness, moderate characteristic assets, and search for and utilize proper advances. It can likewise empower them to differentiate creation for nearby or local markets. The more drawn out the land is beneficial, the more drawn out the benefits will proceed.

Successful farmer-entrepreneurs focus on securing profits, but they are also aware of the long-term negative effect of unsustainable land use. Appropriate land management, in turn, ensures that profits are maintained over the long-term and the farm business is sustainable. A true entrepreneurial spirit should lead to natural resource sustainability because it implies the wise use of scarce natural and human resources in order to stay in business for a long time.

Successful agriculturist business visionaries center on anchoring benefits, yet they are likewise mindful of the long-haul negative impact of unsustainable land utilize. The suitable land administration thus guarantees that benefits are kept up finished the long haul and the homestead business is manageable.

A genuine entrepreneurial soul should prompt common asset maintainability since it suggests perceptive utilization of rare regular and HR with a specific end goal to remain in business for quite a while (Figure 2.3).

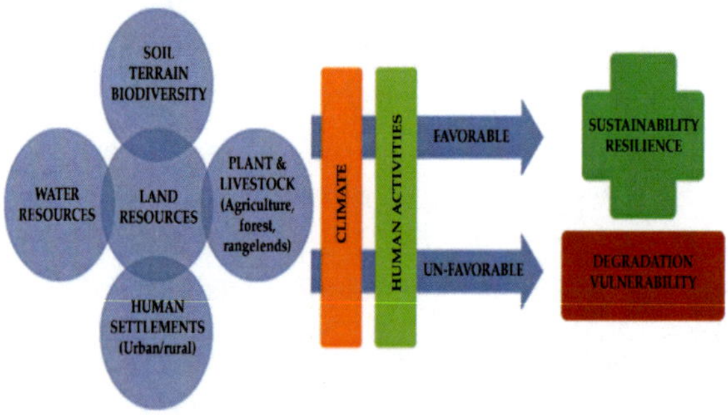

Figure 2.3. Sustainable land management.

Source: http://www.fao.org/land-water/land/sustainable-land-management/en/

2.3.4. Promotion of Group Entrepreneurship

It is possible to cultivate an entrepreneurial spirit among a group of willing participants. It starts with finding a group of farmers who show interest in being self-employed, taking calculated risks and the willingness to work together on a joint production activity.

Forming group business enterprises often have to be initiated from outside the group, for example by extension workers. Until a group becomes united, its members are likely to put their individual interests first.

It is conceivable to develop an entrepreneurial soul among a gathering of willing members. It begins with finding a gathering of agriculturists who demonstrate enthusiasm for acting naturally utilized, going for calculated risks and the readiness to cooperate on a joint generation movement.

Forming group business undertakings regularly must be started from outside the gathering, for instance by augmentation laborers. Until the point when a gathering winds up joined, its individuals are probably going to put their individual advantages first.

Gathering undertakings are regularly defenseless regarding the instruction, proficiency, and administrative experience of its individuals. Such gatherings should be bolstered more profoundly finished a more extended period than would singular farmer business visionaries.

Expansion laborers are likewise expected to make linkages with input providers and markets while building up the administration and entrepreneurial aptitudes of the gathering and its individuals. The connection between the augmentation laborer and the gathering is basic. Reliable help assembles trust. This manufactures eagerness and a more noteworthy soul of solidarity among amasses individuals. The expansion laborer has to know when to quit supporting the gathering.

This ought to happen when the gathering assumes full liability for the administration of its business, however simply in the wake of preparing has likewise been finished. This is a test of expansion specialists and augmentation associations. On the off chance that help is halted too early, the gathering's structure may fall under the heap that is put on it before it is prepared.

On the off chance that it carries on too long, it makes more noteworthy reliance and the move towards confidence is impeded or even ceased. The procedure of gathering business enterprise improvement is effective when it results in a reasonable wander – an agreeable, a rancher affiliation or different types of joint effort (Figure 2.4).

Figure 2.4. Group entrepreneurship.

Source: http://www.thetailoredgrp.com/the-tailored-group-explore-the-mindset-of-successful-entrepreneurship/

2.4. STRATEGIES FOR ESTABLISHING AGRICULTURAL BUSINESS

Notwithstanding procedures to build pay, some of the time it is best to balance out the farmer business. Balancing out systems includes different methods for keeping up the size and association of the homestead business at its present level. These methodologies are embraced from the beginning of an endeavor or after extensive stretches of development, contingent upon the objectives and yearnings of the agriculturist. Conceivable methodologies include no change, full business, satisfactory wage, benefit, and 'keep a watch out.'

- **No Change:** The present cultivating business meets the objectives of the agriculturist.
- **Full Employment:** The business is extended until to the point that the relatives are completely utilized. Now development stops and is kept up at this level as long as everybody is completely locked in. Full-time procured work to grow additionally isn't viewed as a choice.
- **Adequate Income:** The business is extended until to the point when it creates sufficient wage for the agriculturist. Notwithstanding, if satisfactory salary levels are not kept up after some time, the business may extend until to the point that the coveted level of pay is come to.
- **Profit:** After a time of development, the farmer may choose to forfeit future development by utilizing the salary created for utilization purposes and not reinvesting benefits into the business.
- **Wait and See:** An agriculturist may think the current financial and business condition is excessively unverifiable for development as of now and choose to hold up until the point that the earth has enhanced before extending. In the event that this is not overseen effectively, an agriculturist may risk holding up too long and missing an open door for an extension.

Strategic administration comprises of key arranging, key usage, and vital control. Strategic administration manages wide issues and openings influencing the homestead business. It is not the same as everyday administration of the homestead business that arrangements the quick operational issues of the business.

2.4.1. Strategic Planning

This includes:

- Selecting the long-haul objectives of the homestead business and its center qualities;
- Analyzing the outside and inside conditions;
- Developing the most ideal way the objectives can be accomplished in its present condition. Arranging a trek is a decent case.

One can pick a goal – yet what is the ideal approach to arrive? Walk, take a transport, and pass by bike? This requires arranging out the trek in light of every probability and choosing which the best one is. A straightforward rule to take after when creating techniques is to exploit the business' inner qualities. A homestead business methodology ought to be based on what it is great at doing.

2.4.2. Key Execution

This is about sorting out the farmer in a way that guarantees that the procedure can be viably actualized. It includes planning structures and methods and acquiring and coordinating the assets expected to put the procedure without hesitation.

2.4.2.1. Vital Control

This includes outlining approaches to screen and control actualizing the arrangement, contrasting real outcomes with the objectives set, observing the business condition and rolling out improvements to the farm structure, execution design or even the system to meet the objectives.

This incorporates:

- Selecting the whole deal goals of the estate business and its middle characteristics;
- Analyzing the outside and inside conditions;
- Developing the best way, the destinations can be refined in its current condition. Arranging a trek is better than the average case.

One can pick an objective – yet what is the best way to deal with? Walk, take a vehicle, and go by bicycle? This requires arranging out the trek in light of each likelihood and picking which is the best one. A clear govern to take after while making procedures is to abuse the business' internal characteristics. A property business philosophy should be founded on what it is extraordinary at doing.

2.4.2.2. Key Execution

This is stressed over dealing with the farm in a way that ensures that the technique can be feasibly completed. It incorporates arranging structures and strategies and getting and planning the advantages anticipated that would put the method decisively.

2.4.3. Fundamental Control

This incorporates delineating ways to deal with screen and control realizing the course of action, standing out genuine results from the destinations set, watching the business condition and taking off enhancements to the farm structure, execution outline or even the framework to meet the goals.

2.5. QUALITIES REQUIRED FOR BEING A SUCCESSFUL AGRICULTURAL ENTREPRENEUR

There is a contrast between cultivating business administration and enterprise. Farm business administration is about better arranging, usage, control, and overseeing hazard. Enterprise is tied in with looking forward – recognizing openings, making a dream of how the business will develop, improving, and going out on a limb. A farmer who thinks about the homestead as a business that can possibly develop and create is a businessperson.

Entrepreneurial farmers look at their homesteads and see approaches to make them more productive; they create thoughts and after that make an interpretation of them energetically. They require self-motivation, persistence, and certainty with a capacity to design and arrange the farm business. These characteristics empower entrepreneurial farmers to search out business openings, conceptualize, and start new business thoughts and guide the homestead business to achieve the objectives set.

Business visionaries require something other than their identity or individual characteristics. They require thoughts, openings, and assets. Information enables farmers to settle on educated decisions. It places them in a superior position to contrast the present works on being utilized and choices.

Farmers acquire information through experience and perception – from tuning in to and gaining from different agriculturists, watching how things are done and after that rehearsing it themselves. Augmentation specialists are another wellspring of information. Data and its correspondence is an imperative part of information creation and aggregation.

Farmers require learning in every one of the key regions of homestead administration: arranging, executing, and controlling. They likewise require data about essential creation, gathering, handling, wholesaling, and retailing and about info supply, money related administrations, transport, bundling, advancement, and warning administrations.

There are nine key entrepreneurial skills for an agriculturist business visionary: activity, desire, centered critical thinking, imaginative reasoning, taking risks, adaptability, and flexibility, relational capacities, systems administration and status to learn. With these abilities, farmers will be more prepared to contend in this new condition and make benefits by exploiting new market openings. These skills can be procured through training, background, and preparing.

Notwithstanding being business people, entrepreneurial farmers should likewise be phenomenal agriculturists. This requires specialized capabilities especially in three zones: overseeing inputs, overseeing the creation and overseeing promoting.

Entrepreneurial and specialized capabilities should be supplemented by administrative abilities in the elements of finding, arranging, sorting out, driving, and controlling. The agriculturist business visionary plays out these capacities in every one of the key regions of the homestead business: overseeing data sources, creation, and highlighting.

Accomplishment as a farmer business visionary gets through the capacity of the agriculturist to join the entrepreneurial, specialized, and administrative capabilities practically speaking. Homestead business administration is about better arranging, usage, control, and overseeing hazard.

Enterprise is tied in with looking forward – recognizing openings, making a dream of how the business will develop, enhancing, and taking risks. Business people have some uncommon characteristics or attribute that set them apart from the normal ranch director.

2.5.1. Knowledge

Business visionaries require something other than their own attributes. They likewise require a scope of skills and capacities that can be learned or created through preparing and experience – with the help of expansion. A fundamental and essential piece of any competency is learning. Learning is a key factor in effective homestead business administration.

Information enables farmers to settle on educated decisions. It places them in a superior position to contrast the present works on being utilized and choices. Levels of instruction (particularly proficiency and numeracy) may affect creating and utilizing information.

Farmers who have less formal training can, in any case, be successful students, secure, and utilize information to make their homesteads more beneficial. They may need to work intimately with expansion specialists and different wellsprings of data to ensure that data is exhibited in a way that they can get it.

A successful entrepreneurial farmer takes charge of his or her own particular learning. Agriculturists get information in various ways. They learn through understanding and perception and from composed, verbal or visual data. A portion of their insight has been passed on from their folks and grandparents. Numerous farmers additionally get their insight from tuning in to and gaining from different agriculturists, watching how things are done and afterward honing it themselves.

Extension laborers are another wellspring of learning. Whatever source they utilize, agriculturists, as the vast majority, learn best through involvement, by doing. Data and its correspondence are a vital part of information creation and aggregation. The issue is that in numerous nations, data is not generally accessible and data frameworks are ineffectively created. Where data is accessible, farmers think that it is difficult to utilize. For instance, market prices are frequently communicated as normal costs for a district or broadly and might be altogether different from nearby costs.

Agriculturists require learning in every one of the key territories of farm administration: arranging, executing, and controlling. They require data about their immediate capacities – essential creation, collecting, handling, wholesaling, and retailing.

They additionally require data about their help capacities – input supply, money related administrations, transport, bundling, advancing, and warning administrations. Ranchers handle learning in unexpected ways.

Agriculturists that are more customary tend to stick to the information they gained from their forefathers.

Market-situated farm business visionaries effectively look for new and dependable data that will enable them to choose how to make their homesteads more beneficial. How farmers handle learning is a decent pointer of how genuine they are tied in with making benefits and being a businessperson.

2.5.2. Entrepreneurial Competencies

There are nine key entrepreneurial skills for an agriculturist business visionary: activity, aspiration, centered critical thinking, imaginative reasoning, taking risks, adaptability, and versatility, relational capacities, systems administration and status to learn. With these skills, agriculturists will be ready to contend in the changing condition and better ready to make benefits by exploiting new market openings.

These capacities can be created and refined through training and experience. They can likewise be fortified through preparing. Activity: Initiative is an eagerness to work. Entrepreneurial agriculturists are anxious to go ahead. They do not sit tight for others to begin. Entrepreneurial agriculturists lead the way and are the first to act when another open-door tag along. They realize what should be done and can make and express their vision for progress.

2.5.3. Interpersonal Abilities

Farmer entrepreneurs comprehend that their prosperity regularly rests in the hands of other individuals. Therefore, they perceive the need to work with others. They are great communicators; they straightforwardly share data, and they effectively tune in to everyone around them. Most importantly, they are straightforward and dependable when working with others and in the entirety of their business dealings.

2.5.4. Entrepreneurial Skills

Opportunity abilities incorporate having the capacity to perceive and acknowledge business openings. They additionally incorporate market and client introduction, a sharp consciousness of dangers, the capacity to advance and hazard administration aptitudes. Methodology aptitudes incorporate having the capacity to create and assess business methodologies.

They likewise incorporate the capacity to get and make utilization of input and to reflect. They include observing and assessment aptitudes, theoretical abilities, vital arranging abilities, key basic leadership aptitudes, and objective setting aptitudes. Reflection incorporates self-information and knowing one's qualities and shortcomings.

Vital arranging includes the capacity to dissect the business circumstance and its aggressive condition. This, thusly, incorporates the capacity to distinguish and follow up on novel offering openings. Collaboration and systems administration aptitudes incorporate the capacity to arrange and use contacts, and in addition correspondence, collaboration, and participation abilities – especially with different agriculturists and organizations. Authority is additionally a vital component of participation and systems administration aptitudes.

2.5.5. Technical Competencies

Notwithstanding being an entrepreneur, entrepreneurial agriculturists are additionally superb farmers. They apply every one of the capabilities of a business visionary to their cultivating exercises. Specialized skills are required especially in three zones: overseeing inputs, overseeing the creation and overseeing promoting. What separates entrepreneurial agriculturists from different farmers in these parts of cultivating is their conduct. The agriculturist business visionary is dependably vigilant for better approaches to get things done and will try.

- **Managing inputs:** Sources of info are a key component in beneficial market-situated cultivating. The agriculturist businessperson is great at recognizing, sourcing, and getting contributions for the homestead. To oversee inputs effectively, the agriculturist has to realize what inputs are required for every venture, where to get them and how to utilize them. The rancher businessperson has this information and aptitudes, but on the other hand, is continually searching for better quality sources of info; bring down info costs and options that are more effective. The rancher business visionary is prepared to test and learn.
- **Managing marketing:** So as to influence benefits, produce must be advertised and sold. Farmer business visionaries know where the most gainful market is for every item. They are great at arranging contracts. They are likewise dependably vigilant for more beneficial markets and, when found, adjust rapidly to

exploit new marketing opportunities.

- **Controlling:** Successful entrepreneurial agriculturists comprehend the significance of controlling the farm business and ensuring that execution goes as indicated by plans and desires. It is here and there additionally called observing. Controlling suggests checking the advance of the farm business – to some degree or all in all after some time.

It isn't simply watching the progress, however giving standard checks as plans are being actualized, regardless of whether it has been executed effectively. It calls for day by day observing of what happens and contrasting real execution and results against desires. This makes it conceivable to recognize and actualize changes in accordance with different parts of the farm business to enhance execution.

An entrepreneur is a key figure in financial advance. He is the individual who presents new things in the economy. He is considered as the business pioneer and not as straightforward proprietor of capital. He is a man with an adjustable staff, drive, and ability who sees business openings and instantly seizes them for abuse.

M.M.P. Akhouri, in the past Executive Director, National Institute for Entrepreneurship and Small Business Development (NIESBUD), New Delhi, depicts business visionary "as a character who consolidates imaginativeness, preparation to go out on a limb, detecting openings, recognizing, and activating potential assets, worries for perfection and who is tenacious in accomplishing the objective." To be effective, a business person ought to have the accompanying trademark highlights.

- Need to accomplish: Entrepreneurs have a powerful urge to accomplish higher objectives. Their internal identity propels their conduct towards high accomplishment: a large portion of the general population long for progress, however, don't make any move towards accomplishing these fantasies. Business people with high n-Ach factor act constantly to accomplish the objective and influence their fantasies to work out as expected. For them, winning is an accomplishment.
- Freedom: Most of the entrepreneurs begin without anyone else in light of the fact that they are hesitant to work for others. They want to work for themselves and need to be in charge of their own choices.
- Hazard bearing: Entrepreneurs are the people who make choices

under vulnerability and hence they will go for risks, however, they never bet with the outcomes. They pick direct hazard as opposed to playing wild bet. They, accordingly, embrace ascertained hazard which is sufficiently high to energize, yet with a genuinely sensible opportunity to win.

- Locus of control: According to Rotter's locus of control hypothesis, an individual sees the result of an occasion as being either inside or past his own control. Business visionaries put stock in their own particular capacity to control the results of their undertaking by affecting their financial condition instead of leaving everything to good fortune.
- Perseverance: Entrepreneur has the nature of adhering to work he chooses to embrace. When focused on a particular objective and game-plan, business visionaries end up retained to it. They actually take care of the issues that go over their way while setting up the undertaking. They likewise work truly until the point when the entire venture is effectively actualized.
- Positive self-idea: Entrepreneurs are constantly positive in their activity. Being an achiever, he coordinates his dreams and dreams towards the accomplishment of advantageous objectives and sets the phenomenal standard of greatness in what he is doing. This depends on his consciousness of SWOT investigation, i.e. his qualities, shortcomings, openings, and dangers. He uses his positive information to help his reasoning. He never shows any negative disposition.
- Capacity to discover and investigate openings: Entrepreneurs are constantly aware of chances. They rush to see and grab opportunities. They display a creative transformation of the brain and change over the issues into practical openings. They design mentally and envision painstakingly how to accomplish their objectives in understanding an opportunity.
- Any expectation of progress: Hope of accomplishment is a noteworthy nature of entrepreneurial identity. Business people set their objectives with a desire for progress instead of the dread of disappointment. This is on account of they set their objectives based on realities and their capacity to move them further bolstering their good fortune.
- Adaptability: Most of the successful entrepreneurs measure the

upsides and downsides of a choice and tend to change if the circumstance requests. They never feel hesitant to re-examine their choices. They are the people with receptive outlook without inflexibility.

- The expository capacity of the brain: Entrepreneurs are unaffected by individual preferences. They remain past these sorts of preferences as they are reasonable in their approach. At the season of their need, they select specialists instead of companions and relatives to help them. They, as a rule, stay away from the passionate and delicate state of mind towards their business or issue.
- Feeling of adequacy: Entrepreneurs are constantly arranged towards activity for the achievement of their objectives. Being certain of their capacities, they end up as issue solvers as opposed to issue avoiders. They chalk out their objectives for future and make wanting to accomplish them.
- Receptiveness to input and gaining as a matter of fact: Successful business people jump at the chance to have a quick criticism of their execution. They change their plans based on the criticism they get from the earth around them. They gain from their experience and never get demoralized having gotten negative data. Despite what might be expected, they are invigorated by troublesome data to include themselves truly in their own errands to achieve their coveted objectives.
- Standing up to vulnerability: Successful entrepreneurs are constantly hopeful and take each odd as the opportunity. They move their condition such that the works get proficient judiciously. In this way, they win through the use of their exceptional understanding and ability.
- Relational abilities: Entrepreneurs are constantly agreeable while managing individuals at all levels. They connect with crude material providers, clients, investors, and so on for various exercises. As fruitful business people, they ought to be people who like working with others having the genuinely necessary nature of relational ability to manage individuals.
- Need to impact others: Once the entrepreneurs set their objectives, they need to assume the parts of the administrator as well. For affecting others (n Power), a low need to build up an enthusiastic

relationship (low n Affiliation), and a high need to train one's own particular self (to restrain over articulation of their identity) are basic.

- Stress takers: Entrepreneurs are equipped for working for extended periods of time and fathoming diverse complexities in the meantime. As the chief of an industry or a venture, a business person faces various issues and in the right minute, he takes the right choices which may include physical and in addition mental pressure. He can confront these difficulties on the off chance that he has the ability to work for extend periods of time and hold himself cool under repetitiveness.
- Time introduction: Entrepreneurs foresee future patterns basing upon their past experience and presentation. They adhere to the time even-minded while doing their employment.
- Pioneers: Successful business visionaries are trend-setters. They always put their endeavors in presenting new items, a new technique for a generation, opening new markets and perceiving the undertaking.
- Business correspondence aptitude: With a specific end goal to propel others in the business visionaries must have great correspondence ability. Both composed and oral relational abilities are vital for the business people for running venture proficiently.
- Adaptive staff: Successful business people constantly tend to think ahead. They have adaptive resources which influence them to think for what's to come. The future introduction makes them very aware of the changing states of the time and they tend to create merchandise and wares according to the evolving requests.
- Leadership: Entrepreneurs ought to have the nature of authority. Authority is the capacity to apply relational impact by methods for correspondence towards the accomplishment of objectives. Business people as the pioneers ought to give the vital start to inspiration by controlling, motivating, helping, and coordinating the individuals from the gathering for the accomplishment of solidarity of activity, endeavors, and reason.
- Subsequently, business visionaries by their own particular administration styles and conduct lessen the issues by appropriate treatment of circumstances. Great managerial work relies on the

powerful initiative of the business visionary.

- Business arranging: Planning infers choosing ahead of time what, when, and how to complete a thing. Business people ought to be furnished with expertise and information to set up their marketable strategy. An effective business person dependably takes after the standards of administration while getting ready for his business. The arranging can go about as an extension between the present positions and expected future state of the endeavor. It gives a feeling of vision to the business visionaries to adapt to unsafe and indeterminate circumstance.
- Basic leadership: Decision-production expertise is a crucial thing for an entrepreneur. This infers the capacity of picking a specific strategy at each phase of formation of a venture out of a few elective courses to achieve determined objectives. Subsequently, basic leadership is essential consistently and generally at states of vulnerability and hazard.
- Capacity to prepare assets: Entrepreneurs must be able to marshal every one of the contributions to acquire the finished result. They need to assemble 6Ms, i.e. Man, Money, Material, Machinery, Market, and Method adequately to realize the last item as business is a component of whole filling and information finishing.
- Fearlessness: Entrepreneurs must have the self-assurance to achieve the undertaking successfully and effectively. They should take choices all alone in indeterminate and unsafe circumstance and should stick to it certainly regardless of whether there happen beginning misfortunes.

Entrepreneurs have some special qualities or characteristics that set them apart from the average farm manager. Some 'typical' characteristics of a farmer- entrepreneur are shown in the Figure 2.5. They can be grouped into six categories: core values, problem-solving, flexibility, drive, competition, and confidence.

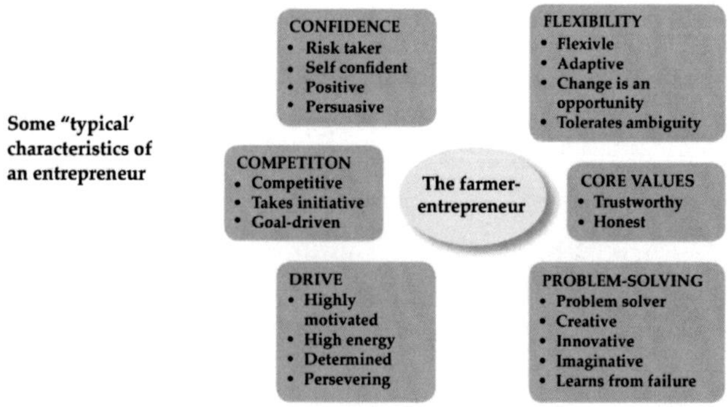

Figure 2.5. Entrepreneurial characteristics.

Source: https://graduatefarmer.co.ke/2018/02/09/typical-characteristics-of-a-farmer-entrepreneur/

It is the attributes appeared in the figure over that empowers entrepreneurial farmers to search out business openings, conceptualize, and start new business thoughts, accumulate the physical, money related, and HR expected to begin the business, set objectives and guide the homestead and all its assets to achieve those objectives.

Not all farmer business people have these attributes to a similar degree. Yet, they will have every one of them to some degree. Without their center estimations of dependability and trustworthiness, their critical thinking nature, their adaptability, their drive, the sense rivalry, and their certainty, they would not by any stretch of the imagination be business visionaries.

All together for good homestead administrators to wind up really entrepreneurial they should build up these attributes.

2.6. CHARACTERISTICS OF SUCCESSFUL AGRI-PRENEURS

Agri-entrepreneurs have individual attributes that set them apart from others and impact the manner in which they work for organizations. Bairwa et al. (2014) portrayed agri-business people as visionary, inquisitive, proactive, decided, diligent, genuine, persevering people with honesty, and administration and association abilities. According to them, agri-business visionaries regularly take part in an assortment of exercises, which recommends that they multi-assignment well. They perceived business people as resolute, driven, aspiring, imaginative issue solvers, yet commonsense and objective arranged. Likewise, they recommended that agri-business visionaries could perceive needs as circumstances. In conclusion, agri-business people energetically acknowledge dangers important to transform thoughts into genuine and attractive items.

Sancho (2010) gave an exhaustive rundown of the various attributes required for being a successful entrepreneur in the agricultural sector. These incorporate activities, feeling of chance or instinct, and autonomy. Agri-business people are likewise seen as unique, showing authority aptitudes, a solid character, and objective driven. Other attributes for the same are a duty, genuineness, committed, enthusiastic, diligent work, and experience.

Sancho (2010) likewise trusted that agri-business visionaries are inventive and creative issue solvers who can transform imaginative thoughts into the real world and hope to prevail in their endeavors. What's more, those engaged with agri-enterprise are proficient and included, and methodically assess their position. In addition, agri-business people esteem associations and collaboration. They promptly go up against assignments, however, stay adaptable and versatile as they go about as change operators.

Thus, Hossain and Jaim (2011) contended that, as proprietors of an association, agri-business visionaries should take an individual stake in the undertaking, including all dangers and vulnerabilities engaged with running it. They fought that agri-entrepreneurial achievement results from firm execution. Anderson et al. (2006) proposed that entrepreneurial aptitudes identified with agribusiness are positive thinking, development, activity, and hazard taking.

They trusted that agri-business people readily acknowledge challenges, assume liability and manage the obscure and that they ought to have the capacity to impart viably, have constructive relational aptitudes, arrange effectively, and lead their business task. Anderson et al. (2006) shared the perspective of Hossain and Jaim (2011) that agri-business people are opportunity searchers and change operators who can examine, orchestrate, assess, design, and execute thoughts.

These characteristics guarantee that the individual/performing artist/operator is pertinent in their entrepreneurial skills under any situation. For instance, entrepreneurial fighters exchanged cigarettes for items not effortlessly acquired amid World War II when purchaser merchandise was proportioned (but these entrepreneurial exercises were here and now answer for item deficiencies and completed with the end of the war). A drive to succeed, a high level of intensity, a high vitality level, receptivity to change, and a valuation for help systems are characteristics of successful female entrepreneurs (Quinay, 2013). Different traits are a positive attitude, confidence, capacity to see new opportunities, the ability to work for extended hours, and acknowledgment of the requirement for continuous business training. Adaptability is vital to female entrepreneurs in struggle zones, as they should be clever in building up their attitudes and plans of action with the goal that they can gain a wage in the event that they are dislodged. Attributes of effective entrepreneurship incorporate a business that survives 3 years (Makhbul and Hasun, 2010), speaks with partners (Islam et al., 2011), sets up informal organizations, increases individual fulfillment (Gorgievski et al., 2011), and balances work with life.

A businessperson's prosperity is judged by their capacity to produce benefit identified with the business' equal the initial investment point (Gorgievski et al., 2011; Miskin and Rose, 2015). Likewise, successful female entrepreneurs see business disappointments as a learning opportunity.

Agri-enterprise training increments conduct aims towards opening a business (Mohamed et al., 2012). Morgan et al. (2010) proposed that agri-entrepreneurial ranges of abilities are circumstance particular. For instance, conflict zone business people should offer items that are effortlessly transported keeping in mind the end goal to gain pay. Past exchange (Section 2.6) of the association between a business visionary and T'boli and Bagobo clans to make unmistakable themes for form purses addresses circumstance particular abilities (Hampel-Milagrosa, 2014). Agri-business people need to feel positive about their capacities to build up an effective business (McElwee and Smith, 2012). They are inside headed to succeed.

Qualities most connected with agri-entrepreneurs (Figure 2.6) incorporate a considerable lot of the terms related to the non-specific enterprise.

Figure 2.6. World cloud of the most common descriptors of agri-entrepreneurs.

Source: https://cab.presswarehouse.com/sites/stylus/resrcs/chapters/1780647751_1stChap.pdf

2.7. OBSTACLES TO POSITIVE AGRI-ENTREPRENEURSHIP CHARACTERISTICS

While a businessperson is close to home, attributes, and experience add to agribusiness achievement, they additionally affect fizzled agribusiness. For instance, hazard avoidance can add to unfruitful agribusinesses (Pannell et al., 2000). Different qualities of unsuccessful agri-entrepreneurs incorporate poor monetary administration, insufficient cost examination, and absence of regard for the item lifecycle offers unfruitful items, unbeneficial ominous buyer discernments, and unmindful client benefit (Duft, 2015).

Moreover, people group, associations, framework, and social capital that help enterprise and employment creation are harmed (Alexander, 2012). Conventional wrongdoing, debasement, political, and arrangement uncertainty likewise upset entrepreneurial achievement. Furthermore, struggle upsets, advertise, and dislodge individuals.

The disappointment of a Somalian slaughterhouse (Massimo et al., 2013) demonstrates that notwithstanding when an opportunity is completed as a business it can fall flat in light of unforeseen dangers. The slaughterhouse was expected to fulfill wellbeing measures for reaped creatures and process neighborhood meat that could be utilized as a substitute for imported items or sent out.

It worked for a not as much as a year prior to shutting. Access to a dissemination framework was an issue as far as on-time meat conveyance. The business did not have an alternate course of action for expanded rivalry in sending out business sectors. In addition, there was a climate issue: an unforeseen dry spell was outside the ability to control the business and contrarily affected collect activities, and additionally salary.

2.8. CASE STUDY: RURAL YOUTH, AGRICULTURE, AND ENTREPRENEURSHIP: A CASE-STUDY OF HUNGARIAN AND ROMANIAN YOUNG VILLAGERS

1. Introduction: Literature characterizes rural areas as areas with a small density of population, high level of unemployment, low productivity, low income of the population, and the very important role of agricultural activity. In the general process of population aging, the topics regarding youth and the role of the young generation in economic activity become very important.

These considerations have led us in choosing the topic of the present paper. The idea of this comparative research emerged from the need to integrate the euro regional outlook and the cross-border approach into the research and development activities within the shared border region of Romania and Hungary.

Based on the comparative research data collected during the cross-border project, our paper focuses on the rural youth as a vulnerable group with difficulties of integration in the labor market.

The main aspect that interests us, however, is the involvement of the young villagers in entrepreneurial activities and local agriculture, viewed as potential solutions both for rural development and for attenuating the rural unemployment of the young population segments by providing them important means of integration into the local and regional economy. The focus of the paper is the rural youth and its involvement in agriculture and entrepreneurship.

One of the concrete practical reasons for initiating this social intervention project and the comparative research, which was part of it, derives from the fact that in both Romania and Hungary youth is the most affected group by unemployment.

In 2014, the specific unemployment rate for the group of young people under the age of 25 was 15.7% in OECD countries, while in the EU member's states it was 23.4%. In Romania, it reached 25.7%1 and in Hungary was as high as 27.2%.

A similar situation also characterizes the shared border region of the two states. In 2011 in Bihor County (Romania), the young people who were at the beginning of their professional careers made up 21.38% of the registered unemployed while in Hajdú-Bihar their proportion was approximately 11% in 2010.4

Being aware of the strong social need to improve our understanding of villages and villagers, in the period of 2012–2013, two partner universities, the University of Oradea in Romania and the University of Debrecen in Hungary, implemented a cross-border project with the aim to promote their integration on the euro-regional labor market (Bihor – Hajdú-Bihar

Euro-region). By means of analyzing socio-economic data and formulating recommendations for public policy decision-makers, the project aimed at contributing to the enhancement of economic competitiveness and the improvement of human resources within the Hungarian–Romanian cross-border area.

Its specific objectives were to offer information, counseling, and orientation and to provide institutionalized training services for the rural youth. The direct target group of the project consisted of 300 youngsters, 150 from each country, and 30 of each of the 10 rural localities.

In the first phase of the project, a comparative cross-border research was implemented, inquiring into the factors influencing the integration process in light of the research data, rural youth can rightly be considered as a vulnerable group, with the meaning of a marginalized group, excluded group, or risk group. All these concepts are in relation to the phenomenon of poverty and social exclusion.

The most frequent meaning of vulnerability is that of weakness, lack of defense, or lack of means. The vulnerable groups are those groups where the support lacks and which are often in poverty and cannot use the opportunities or defend themselves in front of the problems they face. Another meaning of vulnerability is the risk that can lead a person to a poverty level, which is under a limit that is acceptable by the society.

In this larger meaning could be found the young graduates on the labor market, the rural youth on the labor market, and the people who migrate for work (Popescu, 2011). To understand why the rural youth of Romania and Hungary has become a vulnerable group, we will first need to discuss, from the perspective of theoretical positions formulated within economics and the sociology of entrepreneurship, how the rural world developed in Romania and Hungary and what have been the main social and economic transformations of the rural society in both countries during socialism and post-socialism.

Building upon this diagnosis, we will then proceed with the analysis of our research data. The first part of the paper presents the reasons for studying this topic and discusses some theoretical aspects such as entrepreneurs and entrepreneurship (Section 2), the Romanian and the Hungarian rural society (sections 3, 4), and the entrepreneurial environment in Romania and Hungary (Section 5). The second part of the paper presents the research methodology and the results (sections 7 to 11). The last section presents some conclusions.

2.8.1. Entrepreneurs and Entrepreneurship

2.8.1.1. Economy and Sociology of Entrepreneurship

The attitude towards risk is a specific feature of people. Few people assume risks when others do not prefer to risk. Between these two types of people, there is, however, a strong similarity, as rightly pointed out by Peter Drucker, one of the main leaders of management science. Those who assume the risk and those who do not assume the risk make two main mistakes throughout a year (Doltu, 2011).

The focus of this part of the paper is an analytical debate concerning those people who decide to assume risks, and therefore we use to call them entrepreneurs. It is very curious that the word "entrepreneurship" is rarely mentioned in the manuals of economics. The word "entrepreneur" is absent from the writings of Adam Smith or David Ricardo. Marx is not focusing on "entrepreneur" either but rather on the process of work, the control and the leadership of the workforce. The first great economist who uses the word "entrepreneur" is the French Richard Cantillon.

In his view, the entrepreneur is a speculator. In the economic thinking of Walras, the employer is the entrepreneur; however, entrepreneurship is not a production factor but rather a function of an agent, for instance, the capitalist or the manager.

As "marginalism" became the basic paradigm of economics, it is not curious that all the current economics manuals are rich in information with regard to the consumer's behavior, profit maximization, the wage theory, or the theory of international trade, but are very poor in approaching topics such as technical change, the development of big businesses, the causes of welfare and poverty of nations, and the theory of entrepreneurship (Swedberg, 2000).

2.8.1.2. Rural Entrepreneurship

Voicu (2006), one of the most gifted Romanian sociologists from the young generation, asked himself what represents the rural world His answer points to "a fascinating world, where eternity has been born for most of us. A world with unwritten stories, with oral traditions, with hens and sheep… a world of difference for most Western Europeans… a poor world by current standards, but with very rich in understandings and meanings" (Voicu & Voicu 2006, translation by authors).

Eugen Crihan expresses a somewhat different opinion attempting to consider the effects of modernization as well as in a recent book. He believes that "the rural is often perceived as the sheer opposite of the urban. There are two major prejudices. The first one is to consider the rural space as a place of historical backwardness (and of course that of conservative-traditionalist mentalities) [...] the second one is the identification of the rural with the peasantry..." (Crihan, 2011, p. 15, translation by authors).

These two prejudices are hidden under the neutral formula of the urbanization level considered as a synthetic indicator of the development level of every social space. Under this formula, rural is considered a socio-economic situation defined by the existence of many peasants. However, this perception is "without object because the backward and conservative peasant is in a process of extinction following the general trend of modernity" (Crihan, 2011, translation by authors).

What definition can we give to rural entrepreneurship? An answer to this question could be "a force which mobilizes other resources for covering an unsatisfied demand of the market" (Doltu, 2011). There are many cases of successful rural entrepreneurship in countries such as the United States of America, Colombia, Paraguay, or Bulgaria.

The diversification of using nonagricultural available resources through the entrepreneurial combinations created diverse forms such as tourism, offering conditions for sport and entertainment or training, small trading, industrial activities, and consultancy or productive activities such as meat, wood, or milk-producing (Doltu, 2011).

2.8.1.3. The Romanian Rural Society

Romania was a predominantly rural society until the coming of communism because around 80% of its inhabitants were living in the countryside. Even today, when the country is a member of the European Union, Romania is the country with the highest percentage of the rural population (around 45%).

It is not by chance, therefore, that Romanian sociology has begun with the study of the rural areas, and thanks to its innovative contributions having its social roots in the Romanian rural reality it has become very well known in the international sociological world. This has become possible due to the prodigious research activity of the Bucharest Sociological School founded by Dimitrie Gusti.

The monographic research initiated by Dimitrie Gusti has been connected with the idea of the cultural development of the rural Romanian world. The research followed by action was in the opinion of Gusti the solution for the socio-economic development of the Romanian villages.

The Romanian inter-war village was not only a place of the idyllic landscapes but also one of poverty, illiteracy, and of the illnesses caused by the misery. Even in these circumstances, at the end of the 1930s, there appeared a social segment of the middle classes known as "chiaburii" (kulaks). Unfortunately, these middle-class villagers were one of the "enemies" targeted by the communist power; they embarked on the forced collectivization of the agriculture, which began in 1949 in the southern and eastern regions of Romania and ended in Transylvania in 1962.

The Hungarian Rural Society Similarly to Romania, the importance accorded to rural research and development in Hungary in the inter-war period arose from the modernization delay of the society as a whole and the needs derivable from this situation. The backwardness of rural areas has been interpreted by social researchers in the context of the overall transformation of society. Ferenc Erdei, a prominent Hungarian sociologist of the period, regarded rural society as an inheritance of the feudal social structure and an obstacle in the process of the country's modernization (Kulcsár, 1983).

He insisted on the need to search for solutions, which are apt to simultaneously ensure the well-being of the village and the preservation of its traditional identity and values. In setting the theoretical and methodological ground for such an endeavor, the idea of Sociologia militants and the monographic method developed within the Bucharest Sociological School led by Dimitrie Gusti have been used as fertile sources of inspiration by Hungarian rural sociologists. Like in all countries of the region, the hopes and aspirations nurtured by the social reformers were interrupted by the centralist projects of modernization initiated by communist governments.

During the 1950s, Hungary, similarly to its neighbors, underwent a Stalinist phase of collectivization, which resulted in a sharp decrease in the rural peasantry's life standards. In the subsequent years, however, under the impact of the popular reactions against the centralist collectivist agricultural system, the Hungarian communist political leadership developed a new model of collectivized agriculture.

The Hungarian agricultural reform was implemented within the New Economic Mechanism. Initiated in 1968, the Reform offered to agricultural managers the opportunity to work in an economic environment, which partially resembled the market conditions: no central plan objectives, a significant degree of autonomous decision-making and internal organization, and competition (Swain, 1992). Among the reform measures, one should mention the increase of agricultural purchase prices, contract sales instead of compulsory deliveries, the higher wages coming closer to those in industry, and with social security benefits (Swain, 1998).

2.8.2. Entrepreneurial Environment in Romania and Hungary

Although today economic development differences between Romania and Hungary persist, stronger links have developed between economic firms, public institutions, civil organizations, and university research entities within the Euro-region and steps have been made for strengthening the social and economic cohesion of the border area.

If we look at the economic map of Romania and Hungary, we can notice that the development gap and the difference in wages, which still exist, are of smaller amplitude and are less visible in the shared border region compared to the average situation of the two countries. In Romania, the western-located counties such as Bihor, Cluj, Timiş, alongside the county of Ilfov, which is in the neighborhood of Bucharest, the capital city, and Constanţa, the main harbor on the Black Sea Coast, are together the most entrepreneurial regions of Romania.

On the opposite side, there are the counties in the eastern and southern parts of Romania such as Ialomiţa, Dâmboviţa, Olt, Teleorman, Vaslui, and Botoşani. While the group of the more entrepreneurial counties is composed of those regions where there are important universities or large minority groups and/or very good conditions for economic activities (capital city, main harbor of the country), the low economic and social development of the second group of counties is probably the result of the absence of big universities and lack of large minorities.

2.9. CONCLUSIONS

The idea of this comparative research emerged from the need to integrate the euro-regional outlook and the cross-border approach into the research and development activities within the shared border region of Romania and Hungary. We believe that in order to maximize its efficiency, integrated development should be based on an in-depth knowledge of socio-economic realities, presupposing a set of indicators and the use of a shared methodology of data collection and interpretation. The research instruments developed within the project and applied in a unitary manner in the selected rural communities at both sides of the border permitted us to obtain relevant and comparable information. Our historical overview points to the fact that the development routes of agriculture and the rural areas in Romania and Hungary present some essentially similar characteristics but also significant differences. Until the epoch of "socialist modernization" in both countries, the rural and agricultural population was the source of living for the great majority of the population, but due to its low productivity, it could ensure only modest living standards for most families.

In spite of the fact that during the inter-war period internationally recognized sociological research initiatives in both countries focused on how to achieve rural and agricultural modernization, no radical breakthrough could be made in this respect before the communist takeover, which – in contrast to the previous gradual reform initiatives – prompted the start of an abrupt and centrally imposed transformation in both societies.

In order to overcome the persisting modernization gap and to consolidate their own social basis, the communist regimes subjected both Romania and Hungary to forced collectivization and industrialization, managing to dislocate a large part of the population from the villages to the cities and towns.

In only a few decades, the proportion of those living in agriculture dropped sharply and the nature of agricultural work changed radically, putting, in fact, an end to the traditional peasant way of living. However, after the suppression of the 1956 anti-communist revolution in Hungary, Hungarian, and Romanian agriculture experienced rather different paths of development.

While in Romania forced collectivization and industrialization continued to aim at the destruction of the spirit of initiative and self-management, which was characteristic to the traditional peasant economy, in Hungary, the communist leadership strived to achieve a kind of new "reconciliation" with the peasantry.

Agricultural cooperatives were granted a large degree of autonomy from the central leadership and were permitted to take economically grounded decisions in order to enhance the individual well-being of their members. After the collapse of the communist regime, re-privatization of the collectivized agricultural farms occurred in both countries, but in rather different conditions and with different consequences, partly derivable from the different realities inherited from the socialist period.

In Romania, where collective farms were rightly perceived as another form of state property and of state exploitation of cheap agricultural workforce, most of them were quickly dissolved and replaced by small individual peasant farms bound by their small size and other economic constraints to produce predominantly for subsistence and self-consumption. In contrast, in Hungary, peasants had the opportunity during socialism to experience the advantages of co-operative work in large farms, and therefore the political chances for preserving this more market-oriented form of labor organization in the post-socialist period were much more favorable than in Romania.

These diverging socio-historical conditions – along and in combination with differences in the level of economic development and modernization – should be regarded, in our view, as key explanatory factors of the results of our research, which show that Hungarian young people involved in agriculture are producing to a larger extent for the market, have more entrepreneurial initiatives, and perceive greater opportunities in their own home region as compared to Romanian young villagers.

Hungarian rural inhabitants live in a comparatively more consolidated socio-economic environment, which was less affected by discontinuity as compared to the Romanian villages. The Romanian rural world lives in a higher degree of economic uncertainty and has to use to a larger extent compensatory modes of earning an income such as seasonal agricultural work or migrant work abroad.

Perhaps the comparatively stronger presence among the researched Romanian young villagers of the intention to start their own business can be partly explained by the comparatively more precarious living conditions that the local agriculture can offer to local rural inhabitants in Romania. Perhaps the most viable solutions would require cluster analyses of rural areas because there are major demographic and socio-economic differences between these zones.

Thus, specific measures need to be taken, adapted to the actual needs and potential of each area (Weingarten, 2004). In this regard, rural typologies can be useful instruments for the design of development policies for rural areas, based on the recognition that the addressed issues are multi-dimensional (Oláh & Pakurár, 2011).

Another possibility emphasized in the specialized literature refers to redefining rural development policies based on a new paradigm of rural development by including the concept of the network (Murdoch, 2000). Murdoch mentions two network categories: the vertical, which connects the rural to the agriculture and food sectors, and the horizontal, which connects rural spaces to the non-agricultural processes generated by economic changes.

Taking into account the important differences between the socio-economic environments in the two countries outlined above, we think that the potential solutions and strategies of development should be adapted as much as possible to the particular conditions of the two countries and to the specific circumstances of each individual rural locality.

In Hungary, efforts should be made in order to encourage local villagers to assume more risks in taking autonomous life decisions and in trying new opportunities and alternative sources of income. In the case of Romanian young rural inhabitants, the stress in counseling and community building activities should probably be laid mainly on persuading villagers to stay in or to return to their home places, and on assisting them to identify and develop new job opportunities (including entrepreneurial and self-employment solutions) at the local level.

In these efforts, stakeholders at local and regional levels should play an essential role. Decentralization of decision-making should be implemented, taking into account that local communities dispose of the best knowledge about the employment needs and opportunities, and the resources within the local community.

Local governments, civil organizations, and individual citizens from the two neighbor countries and regions, including the young people themselves, should learn from one another and establish links of co-operation. The data from our quantitative research convinced us that a qualitative approach that would refine the quantitative data would be necessary in the near future or a next research project.

It would be interesting to know the opinions of different local leaders from the villages in the Bihor–Hajdú-Bihar Euro-region regarding the problems such as agriculture, co-operative-farms, individual farms, and entrepreneurship and future opportunities for the rural youth. The research also raises questions regarding the linkage between the economic environments of the region and entrepreneurial opportunities and activities. Maybe a future research of our team will be dedicated to this topic.

REFERENCES

1. Kahan, D., (2012). *Entrepreneurship in Farming*. Food and Agriculture Organization of the United Nations. Rome.
2. Olah, Ş., & Flora, G., (2015). *Rural Youth, Agriculture, and Entrepreneurship: A Case-Study of Hungarian and Romanian Young Villagers*. Retrieved from http://www.acta.sapientia.ro/acta-econ/C3/econ3–03.pdf.
3. Raj, S., (2018). *Entrepreneurial Characteristics*. http://www.skynetplanet.com/entrepreneurial-characteristics/?i=2.
4. Sullivan, P (2017). *Agri Entrepreneurs and Their Characteristics*. Nashville (Tennessee) https://cab.presswarehouse.com/sites/stylus/resrcs/chapters/1780647751_1stChap.pdf
5. *Typical Characteristics of a Farmer Entrepreneur*, (2018). https://graduatefarmer.co.ke/2018/02/09/typical-characteristics-of-a-farmer-entrepreneur/.

CHAPTER 3

AGRICULTURAL ENTREPRENEURSHIP AND DEVELOPMENT

CONTENTS

The chapter displays new research that has exuded from the entrepreneurship domain. It largely centers on the firm and the individual business people, investigating portfolio enterprise inside the ranch part (cultivate based pluriactivity), the precursors and results of different business proprietorship, its asset prerequisites, and the impact and impacts of the homestead family unit. This part gives confirmation of the heterogeneity of entrepreneurial exercises in the ranch division. It likewise examines pull factors versus push factors as inspiration for new business new companies.

3.1. INTRODUCTION

The horticultural division has encountered significant changes as of late. Strategy changes have lessened the size of help in agribusiness, while request side changes, in the type of expanding retailer focus and complex examples of purchaser inclinations, have required the advancement of modern market-driven procedures.

Agriculturists have exhibited uneven capacities to adjust and conform to these continuous changes. The capacity and affinity of ranchers to take part in entrepreneurial practices is a key clarification of various examples of reactions inside the division.

The progressing changes that have happened inside the agribusiness segment have pulled in consideration from a wide range of orders (Figure 3.1).

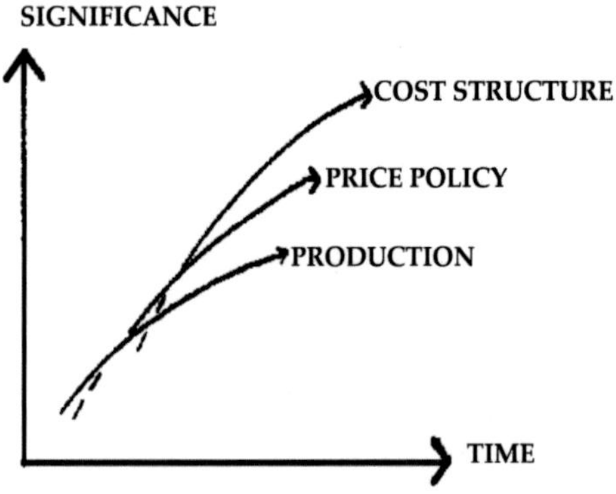

Figure 3.1. Agricultural development through entrepreneurship.

Source: http://www.fao.org/docrep/W6882e/w6882e02.htm

Rural financial specialists and rustic sociologists have reported the effect of these progressions on the cultivating area and their encompassing networks.

Whereas, business enterprise researchers have concentrated on the capacities of a few ranchers to perceive and assess new openings and to utilize cultivate based assets in the misuse of new, regularly non-rural markets.

The outcomes have been the improvement of two separate disciplinary methodologies, with particular starting points, standards, and strategies, both of which are worried about the part of new pursuit creation in-country financial rebuilding, yet with little confirmation of any linkage between the two methodologies.

Farming is not the most common of areas for enterprise scientists to ponder; their spotlight is customarily prepared on rising innovation-based ventures that can exhibit abnormal amounts of advancement and the guarantee of future riches creation and business development.

Nor is agribusiness as an area known for large amounts of enterprise; for most investigators the characterizing normal for cultivating is the abnormal state of help given to the division as sponsorships, value bolster and other market administrative instruments, where the 'essential factor for their prosperity isn't business insight to such an extent as their viability as political entryway.

However, solid connections exist between the farming and the enterprise investigates areas. The horticulture division is made out of little scale, family-possessed or rented ranches that have largely survived the progress through ages. Regardless of expectations despite what might be expected, rebuilding, and the development of vertical joining inside the sustenance and agri-business ventures have not prompted the end of the little ranch.

That the area has survived and flourished, in spite of the changes of the previous decades, is largely inferable from the entrepreneurial limits of ranchers and agriculturalists.

3.2. THE EXCLUSION OF AGRICULTURE FROM ENTREPRENEURSHIP RESEARCH

While a few peruses may address why enterprise researchers should think about agribusiness, an even more intriguing and applicable inquiry to posture is "the reason has the business enterprise explore area overlooked the horticulture division for so long?' The confirmation exhibiting the rejection of farming from a business look into is overpowering. The sectoral focal point of enterprise inquire about has for the most part been fabricating; high innovation and administration businesses.

The concentration is effortlessly legitimized; to shifting degrees, these divisions show high rates of new firm arrangement and work development potential.

By differentiate; agribusiness is described by a decrease in work and foundation numbers, high hindrances to business start-up and complex market administrative components. Not exclusively is the farming part perceptible by its nonattendance from the enterprise look into space, it is frequently clearly avoided in numerous administrative and secretly held private venture datasets.

In the UK, for instance, numerous administrative information sources, for example, the Census of Production, particularly bar agribusiness, while others, for example, the Census of Employment do not gather information from rural or agricultural organizations.

In the USA, the yearly province of Small Business Report to the President gives an account of a business populace that contains 'non-cultivate sole proprietorships.'

The detachment of farming from different types of big business inside generally utilized informational collections postures handy troubles for specialists who may, as a result, unquestioningly accept the presence of a level of rural specificity that refutes examinations with other (non-agrarian) kinds of big business.

While the avoidance of farming from business enterprise might be experimentally advantageous, it is hypothetically confounding. A few of the establishment figures of enterprise hypothesis.

For example, Richard Cantillon, Jean-Baptiste Say, and John Stuart Mill were composing when horticulture was the overwhelming business and the cases they used to delineate their thoughts concerning entrepreneurial rents were ordinarily gotten from appropriation.

Cantillon's (1755) portrayal of a business visionary as 'An operator who contracts with providers at known costs to deliver products that could be sold later at indeterminate costs,' depicts splendidly the vulnerability bearing exercises of eighteenth-century agriculturists who purchased seeds at a known cost so as to develop create which was later sold for an obscure return. Henceforth, the inceptions of contemporary business enterprise speculating are construct to an impressive degree upon eighteenth-century horticultural practices.

In the event that it is eminent that the business enterprise look into the area has expectedly barred in agribusiness, it is similarly remarkable that the spaces of farming financial matters.

It also includes provincial humanism have been moderate in valuing the emotional advancements in the grant that include occurred inside the business enterprise explore space.

To some degree this can be clarified by customarily held perspectives of rural specificity seen as an 'uncommon case,' conventional cultivating was occasionally observed as a business not to mention an entrepreneurial movement.

The production of Gasson et al's (1998) original audit of an edge as privately owned companies was a defining moment in how horticultural endeavors were seen by a few researchers.

Gasson and her associates contended that the tripartite structure of agrarian relations has been destroyed with the decrease of landowners and workers, leaving a leftover mass of ranches, as much as 90 for every penny of which were family possessed or rented and worked, for the most part, by family work.

Despite the fact that Hill (1993) contended that there were risks in expecting that 'little' homesteads and 'family' ranches were synonymous, it was promptly clear that the little family cultivate endeavor has much just the same as non-cultivating undertakings.

Family cultivates were described by an absence of detachment amongst administration and control; benefit expansion goals tempered by the requirement for freedom and progression; choice procedures including various relatives.

The utilization of the family as a wellspring of work and hazard capital; and the nearness of a great philosophy of autonomy, independence, and custom (Gasson et al., 1988).

All through the 1990s, the effect of agro-approach changes concentrated consideration both on an assortment of ranch based exercises that could substitute declining earnings from agribusiness and on the homestead family unit as a provider of work, assets, and thoughts for new endeavors and exercises (Fuller, 1990; Shucksmith and Herrmann, 2002).

While much research has thought about the inspirations, exercises, and execution of new ranch based non-horticultural endeavors, the move towards 'post productivism' and the ascent of multi-practical agribusiness (Evans et al., 2002; Potter and Tilzey, 2005).

The examination of homestead-based business wandering has been directed largely without reference to the enterprise explores space.

3.3. DEVELOPING SYNERGIES BETWEEN ENTREPRENEURSHIP AND AGRICULTURE

Clearly, entrepreneurship has always been an integral aspect of the agricultural sector, a fact often obscured by both the presence of complex market regulatory mechanisms and the extraordinary persistence of the view that agriculture is a special case to be considered separately from other forms of economic activity.

Though the development is, scholarly specialization has meant that there has been little cross-fertilization to date between the domains of entrepreneurship and those of agricultural economics and rural sociology, there are strong arguments that this should change.

The profound changes experienced by the agriculture sector as a consequence of policy liberalization and changes in the demand side and, in particular, the recognition of farmers' uneven propensity to adjust and adapt' has drawn attention to the need to better understand variations in individual attitudes and farm strategies.

Despite analyzes of the shift towards a 'post-productivism' research orientation (Evans et al., 2002), the need to understand the underlying processes that drive farm household strategies is paramount.

In this regard, the domain of entrepreneurship provides considerable research experience and depth of understanding that can be profitably applied to the analysis of the farming sector.

At the same time, the entrepreneurship research domain has much that it can learn from the agriculture sector and from the agricultural research domain. The small family farm enterprise has much in common with the non-farm enterprise and these similarities are likely to continue over time as pressures grow on farm incomes.

As such, the sector offers entrepreneurship researchers a unique opportunity to analyze issues that are central research themes within the domain, such as the family-business nexus, the identification.

It includes exploitation of opportunities, the management of portfolio entrepreneurship, and the role of the family in business performance and growth and business ownership over multiple generations.

The motivation for this edited collection of research on entrepreneurship in agriculture and rural development is to better understand the synergies that may exist between the two subject domains and to explore the fruitfulness of further studies.

The purpose is to both explore the contribution that the entrepreneurship research domain may offer to the debates surrounding agricultural and rural development and to consider whether and in what ways a consideration of the issues surrounding the agriculture sector can illuminate some of the less well understood aspects of entrepreneurship theory and practice.

3.4. THE SEPARATION OF AGRICULTURE AND INDUSTRY

The faction amongst farming and industry is considerably more extensive than is clear inside the business area.

Inside the fields of human science, financial matters, history, and geology, rural creation, and country condition in which it happens is generally viewed as independent subjects of research and instructing.

While the business writing has focused on the mechanical undertaking to the avoidance of farming, different controls have considered the agribusiness part in awesome profundity, yet as a different and particular field of inquiry from modern creation.

The partition of agribusiness from industry is not just a result of the subject specialization of researchers; the division exists inside the monetary, political, and social establishments that encompass the segment.

Overall, these organizations mirror a split amongst horticulture and other financial exercises that fortify a perspective of agrarian uniqueness and specificity.

The superseding requirement for nourishment security as an issue of national and worldwide significance, combined with the capacity of ground-breaking vested parties to activate political help, guarantee that agribusiness has frequently been viewed as an 'extraordinary case.'

At the financial level, this is showed in the help that has been given to the area as creation and value bolster and other market administrative systems.

The post-war introduction towards farming creation, initially presented as methods for guaranteeing the security of sustenance supply, is profoundly dug inside the part.

Farming is additionally portrayed by the presence of particular political organizations. A typical component of both created and creating economies is the presence of committed legislative establishments in charge of managing help to and directing the proficiency working of all farming and agribusiness related ventures.

At the social level, agribusiness is likewise made particular by issues encompassing area residency. The land is recognized from different types of capital in as much as it is settled supply is liable to geographical and climatic requirements and, in particular, is complicatedly associated with a connection structure of possession and stewardship.

In spite of the fact that there is a discussion about the causes of the division of horticulture and industry, plainly it is anything but an ongoing marvel. Kautsky's (1998) investigation.

The Agrarian Question, first distributed in 1899, portrayed the partition amongst farming and industry as rising in the early medieval period. These social orders were independent, steady, and made as a rule out of worker experts.

The presence of little businesses began a procedure of commoditization of horticultural items and spiraling specialization, which at last brought about the detachment of the two types of generation.

While an itemized investigation of the birthplaces of partition is past the extent of this presentation, The Agrarian Question is especially critical for different reasons. Kautsky's depiction of the scope of exercises completed by medieval workers presents us with an early case of what is currently known as ranch expansion, and his examination of the procedure of laborer proletarianization.

The mix of little-scale agrarian creation with off-cultivate business gives a hypothetical system for the investigation of what is currently known as homestead based pluriactivity (Bryden et al., 1992).

In spite of the fact that it is enticing to see farming and industry as particular financial circles, this gives a false representation of the multifaceted nature of the connection between the two types of generation and these verifiable connections amongst provincial and urban social orders. While contemporary investigators regularly see farming as having just a negligible status in connection to industry, verifiably horticulture was viewed as focal, as far as business and profitability as well as in its part in the advancement of industry and urbanization.

As Adam Smith (1776) depicted in The Wealth of Nations, 'The immense trade of each acculturated society is that carried on between the occupants of the town and those of the nation.

The nation supplies the town with the methods for subsistence, and the materials for production. The town reimburses this by sending back a piece of the fabricated create to the tenants of the nation.'

If surplus in horticultural profitability realized the industrialization of the eighteenth and nineteenth century, it was likewise in charge of achieving the expansion in urban residency.

There are various agriculturists and entrepreneurs who believed that the agribusiness and industry are partitioned and are neglected which provides a contemporary view to the agricultural sector from their viewpoints.

The impacts of these progressions have not been, be that as it may, equally spread all through the division and current horticulture is as much described by its decent variety as it is by its unpredictability.

To an extensive degree, the premise of farming 'specificity' has since a long time ago dissolved. To be sure, a quarter century prior country's sociologist contended that, 'The response to the inquiry, what is specific to farming under free enterprise, is nothing.

It is not that farming has created parallel to industry, however, that connections in complex chains interface generation of particular plants and creatures to similarly particular assembling procedures' (Friedmann, 1986: 44).

Given the significant changes that include occurred inside the farming division in the interceding time frame, it is sketchy with reference to whether the detachment of agribusiness industry still exists anyplace past the domain of scholarly subject areas.

3.5. THE FOCUS ON ENTREPRENEURIAL OPPORTUNITIES AND RESOURCES

Some as containing three unmistakable however covering sees (Alsos, 2007) have depicted the business enterprise field. The development-based point of view is from crafted by Schumpeter, who saw enterprise as the making of an unevenness in the market in light of another mix of assets, new markets as well as new hierarchical arrangements.

The business arrangement point of view sees enterprise as the way toward making new business associations, from entrepreneurial goal to the improvement and foundation of new associations.

This procedure of authoritative creation is characterizing as associations rise (Gartner, 1993) or the entrepreneurial procedure. The open door-based point of view puts the quest for an open door at the center of a business enterprise (Shane and Venkataraman, 2000).

Concentrates normally take as a beginning stage Kirzner's (1973; 1997) perspective of the 'ready business visionary' who distinguishes business open doors as defects in the market and organizes assets to abuse these openings, accordingly reestablishing balance (Landstrom, 1999).

In this point of view, the enterprise is characterized by the revelation and abuse of business openings. The open door-based perspective of enterprise shows up especially pertinent inside the setting of the farming segment, suggesting a perspective of business as being worried about the formation of new business exercises through the recognizable proof and misuse of chances of which the arranging of assets constitutes an essential part (Alsos, 2007).

The emphasis on new business exercises suggests that the aftereffects of the entrepreneurial procedure might be another firm or another business movement inside a current firm.

While ranches are, obviously, organizations in their own right, the emphasis on research and strategy consideration has been on the capacity of agriculturists to produce extra openings, sorted out either as new business wanders or as a major aspect of the current business substance.

The idea of rural pluriactivity, portraying ranchers' commitment in the wage creating exercises notwithstanding 'customary' farming generation, is reflected by the parallel idea of portfolio enterprise, the responsibility for organizations, considered inside the business area.

Portfolio business visionaries have been recognized as an imperative 'sort' of businessperson, more aware of business openings and more equipped for viable open door looking.

They are thought to put more accentuation on imaginativeness, have more individualistic and objective situated individual states of mind and are thought to be more entrepreneurial' than different sorts of the entrepreneur

(Westhead and Wright, 1998). One of the fundamental riddles that examination identifying with entrepreneurial open doors has managed is the reason, when, and how a few people create openings while others do not (Shane and Venkataraman, 2000).

There is by all accounts assertion that a few people are ready to distinguish specific open doors than others, and that open-door revelation requests the ownership of the fundamental data also the psychological capacity to assess this data (Gaglio and Katz, 2001; Corbett, 2007).

Openings shift largely in unpredictability and attributes, thus do the procedure through which they are distinguished. There seems, by all accounts, to be no less than three measurements of such varieties.

Right off the bat, either opening might be viewed as a Schumpeterian or a Kirznerian write (Shane, 2003). For Schumpeter ([1934] 1962), business people utilized new data, an aftereffect of research or the age of new learning, to recombine assets into more profitable structures.

Interestingly, Kirzner (1997) contended that open doors exist because of contrasts in access to existing data.

Learning about business sectors, how to serve markets and attention to client issues are especially persuasive variables deciding a person's readiness to circumstances. Besides, openings may emerge fortunately or through consider looking.

The view that business people recognize open doors because of predominant data handling capacity, seek methods or examining conduct places interest in data at the focal point of chance revelation.

Thirdly, there is likewise banter with respect to whether openings impartially exist and should be found or whether openings are subjectively made and molded by the business visionary. The distinguishing proof of chances has been all the more as of late saw as a procedure which included components of both 'target' acknowledgment and 'subjective' creation.

While opportunity recognizable proof and creation is the beginning stage of the entrepreneurial procedure, misuse of the open door is the following stage.

Effective open-door abuse is generally reliant upon the person's capacity to get to or prepare the vital assets to convey the plan to fulfillment. Entrepreneurial assets are generally characterized as the material and unimportant resources controlled by the firm. Firm assets are by and large observed as including the greater part of the benefits, abilities, hierarchical procedures, firm qualities, data, and learning controlled by the firm which makes the firm ready to produce and execute systems that give it the upper hand.

Barney (1991) ordered resourced in three classes: physical capital assets, including geographic areas, premises, hardware, physical innovation and access to crude materials; human capital assets including preparing, knowledge, judgment, knowledge, connections, and experiences of the people.

In addition, authoritative capital assets, including the company's structures and frameworks, and additionally casual relations among bunches inside the firm and between the firm and the earth. Alvarez and Barney (2000) added entrepreneurial abilities to this rundown of asset classes.

Entrepreneurial resources, including learning, information, and innovativeness are immaterial, however, can be viewed as the capacity to perceive which assets are or will be significant, obtain, and grow such assets, join assets in profitable packages and convey them to the errands where they will be most successfully utilized.

As Penrose (1959) proposed, the presence of profitable assets alone is not adequate to make the upper hand – how these assets are utilized is likewise essential.

If there should be an occurrence of portfolio business people or pluriactive agriculturists, their current organizations may fill in as a seedbed for the advancement of new business exercises (Carter, 1998; Alsos, 2007).

Portfolio business visionaries can begin to grow new business exercises on the assets controlled inside their current firms, which may diminish the difficulties identified with the development of an underlying asset base.

A current firm gives access to assets, for example, authoritative schedules, representatives, providers, and clients, and in addition physical assets, for example, structures, and hardware (Alsos and Carter, 2006).

This may give portfolio business people a significant favorable position over fledgling business visionaries who presently cannot seem to gain the fundamental assets to begin in business.

Beginning another business from a current business base may decrease both the peril of the new pursuit and furthermore, by sharing assets between the business exercises, the expenses of another start-up. Studies that have concentrated on asset sharing between business exercises inside a business visionary's portfolio have normally accepted that advantages would spill out of this kind of asset exchange. Verifiable in this approach is the view that assets are seen as resources fit for adding to upper hand.

Even more as of late, it has been perceived that some entrepreneurial assets might be broken, exorbitant, and constitute a risk for the new venture.

Broken and wrong assets are not just obvious inside the physical and authoritative capital assets; they may likewise be available inside human capital assets.

Experienced business people beginning new business exercises may likewise confront liabilities coming about because of their current firms. Starr and Bygrave (1992), for instance, highlighted the setting reliance on entrepreneurial start-up information.

Furthermore, liabilities of experience may come from stagnated systems, an unwillingness to learn, settled capital and different assets and lessened time accessibility. Reuber and Fischer (1993: 31) contended that specific assets could be both resource and risk all the while.

For instance, a long residency in a specific industry may empower a person to perform even more productively or viably when beginning another wander in that industry, yet additionally may repress that person from seeing new openings or choices.

The examination of entrepreneurial activities from a viewpoint that features opportunity recognizable proof and investigation, and the assets accessible to the businessperson, has been an especially productive line of research inside the non-cultivating parts.

The division of horticulture from different types of industry has guaranteed that the investigation of ranch organizations has occurred in a different scholastic circle, and has neglected to go to the consideration of researchers comfortable with speculations, ideal models and methodologies that have been set up inside the space of business enterprise look into.

We trust that the business enterprise inquires about space has an unmistakable commitment to make to the comprehension of the present changes that are occurring inside farming and country zones.

3.6. INNOVATIONS IN FOOD PRODUCTION AND DISTRIBUTION

An exploration takes a division and industry point of view, investigating new advancements in food creation and dissemination frameworks.

These incorporate imaginative activities in production network linkages, vital partnerships and vertical arrangement, and novel reactions to changing customer inclinations and developing familiarity with the significance of provincial marking.

The farming division is hallmarked by a high level of disguise as far as worldwide markets, an expanding number of agro-processing plants and vast multinational firms and a substantial number of little makers, helpful sorting out, and national, territorial, and neighborhood markets.

To put it plainly, there is a substantial assorted variety in food production and dispersion. The diversity might give the fundamental dynamism to development within the sector. Advancement is regularly comprehended in Schumpeterian terms, as new blends of creation factors.

For example, generation of new products, presentation of new procedures, the opening of new markets, access to new wellsprings of crude materials and middle people, and industry association (Schumpeter, [934]. 1962).

This chapter exhibits how effective business people are currently captivating in instituting nature around them, for instance, via looking their systems for thoughts to enhance their organizations and always attempting to build up their plans of action that is, working with incremental advancements.

3.7. AGRICULTURAL ENTREPRENEURSHIP AND ITS SPATIAL CONTEXT

From a financial and strategy point of view, vital sub-subject concerns the connection between economic diversity inside areas and economic development. Much research has contemplated these inside urban settings, however, country settings have been ignored.

There has been an examination occurred in the matter of what degree financial decent variety is related with pay levels in rustic districts. The creator shows that the more assorted a country economy, the higher its pay levels, hence attracting regard for the way that provincial districts need to encourage rustic area expansion rather than specialization.

New farm-based endeavors may make an essential commitment to rustic monetary and social improvement on the encouraging or obliging capital inside the territory. Psychological and culture-based guidelines and shared implications can limit spatial settings. In such manner, neo-endogenous hypothesis places that provincial improvement needs both endogenous and additional territorial assets such as in-vagrants.

They convey new data and information to a country region, in this manner opening new roads for business enterprise and reviving provincial networks if they figure out how to wind up installed in their new networks, while neighborhood proprietors keep up nearby social and social capital.

Community entrepreneurship is another development option for rural regions. It happens collectively and in local contexts, highlighting social commitment, non-profit goals, and benefits for the wider community as (additional) driver for entrepreneurship.

Entrepreneurship then is the advantage for social change as has been illustrated for 'depleted' urban communities. Community entrepreneurship also can foster rural entrepreneurship and development.

An example of community entrepreneurship in the forestry sector, initiated in the early 1990s by public authorities in a Canadian Aboriginal community.

The authors show how 'transient' community entrepreneurship helped to facilitate entrepreneurial initiatives in the forestry sector, demonstrating the usefulness of considering community entrepreneurship as a trigger for socio-economic development in rural settings.

Community entrepreneurship is another advancement choice for rural areas. It happens overall and in neighborhood settings, featuring social responsibility, non-benefit objectives, and advantages for the more extensive network as (extra) driver for business enterprise.

Entrepreneurship at that point is the use for social change as has been delineated for 'drained' urban networks. Network business enterprise additionally can cultivate rustic business and improvement.

A case of community entrepreneurship in the forestry service area started in the mid–1990s by open experts in a Canadian Aboriginal people group.

The creators indicate how 'transient' community entrepreneurship encouraged entrepreneurial activities in the forestry sector, exhibiting the handiness of considering community entrepreneurship as a trigger for financial improvement in rural settings.

3.8. WHAT CAN AGRICULTURE DO FOR DEVELOPMENT

Agribusiness has highlights that make it a novel instrument for advancement. Farming can work working together with different divisions to deliver speedier development, decrease neediness, and support the earth. In this Report, agribusiness comprises of harvests, domesticated animals, agroforestry, and aquaculture. It does exclude ranger service and business catch fisheries since they require limitlessly extraordinary investigations.

However, connections amongst agribusiness and ranger service are considered in the talks of deforestation, environmental change, and natural administrations.

Horticulture adds to improvement from numerous points of view. Horticulture adds to improvement as a monetary movement, as an occupation, and as a supplier of ecological administrations, making the part a remarkable instrument for advancement.

As a monetary action: Horticulture can be a wellspring of development for the national economy, a supplier of speculation open doors for the private part, and a prime driver of agribusiness related enterprises and the provincial nonfarm economy.

66% of the world's horticultural esteem included is made in creating nations. In the agri-based business nation, it creates by and large 29% of the GDP and utilizes 65% of the work compelling. The businesses and administrations connected to agribusiness in esteem chains regularly represent in excess of 30% of GDP in changing and urbanized nations.

Rural generation is critical for sustenance security since it is a wellspring of wage for most of the country poor. It is especially basic in twelve nations of Sub-Saharan Africa, with a consolidated populace of around 200 million and with profoundly factor residential generation, restricted tradability of sustenance staples, and remote trade imperatives in meeting their nourishment need through imports.

These nations are presented for intermittent nourishment crises and the vulnerabilities of sustenance help, and for them, expanding, and balancing out household generation is fundamental for nourishment security.

As a business: Horticulture is a wellspring of vocations for an expected 86% of provincial individuals. It gives occupations to 1.3 billion smallholders and landless specialists, "cultivate financed social welfare" when there are urban stuns, and an establishment for suitable rustic networks.

Of the building up the world's 5.5 billion individuals, 3 billion live in rustic territories, almost 50% of mankind. Of these country occupants, an expected 2.5 billion are in families associated with agribusiness, and 1.5 billion are in smallholder families.

The ongoing decrease in the $1-a-day destitution rate in creating nations—from 28% in 1993 to 22% in 2002—has been for the most part the aftereffect of falling rustic neediness (from 37% to 29%) while the urban neediness rate remained about steady (at 13%).

In excess of 80% of the decrease in the country, destitution is owing to better conditions in rustic regions as opposed to out-relocation of poor people. In this way, as opposed to regular discernments, relocation to urban areas has not been the principle instrument for rustic (and world) destitution decrease.

In any case, the vast decrease in the quantity of rustic poor (from 1,036 million out of 1993 to 883 million out of 2003) has been bound to East Asia and the Pacific (Figure 3.2). In South Asia and Sub-Saharan Africa, the quantity of rustic poor has kept on rising and will probably surpass the quantity of urban poor until 2040. In these districts, a high need is to prepare horticulture for neediness lessening (Figure 3.2).

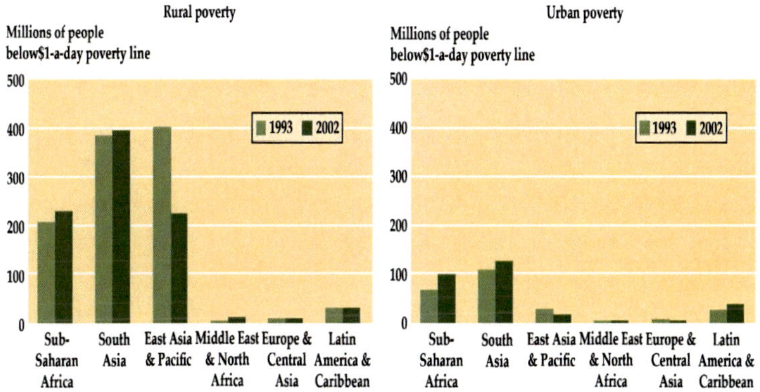

Figure 3.2. The number of poor rose in South Asia and Sub-Saharan Africa from 1993 to 2002 ($1-a-day poverty line).

Source: Ravallion, Chen, Sangraulla 2007

As a supplier of ecological administrations: In utilizing (and oftentimes abusing) normal assets, horticulture can make great and terrible natural results. It is by a long shot the biggest client of water, adding to water shortage.

It is a noteworthy player in underground water consumption, agrochemical contamination, soil weariness, and worldwide environmental change, representing up to 30% of ozone-depleting substance emanations.

In any case, it is likewise a noteworthy supplier of natural administrations, by, and large unrecognized and unremunerated, sequestering carbon, overseeing watersheds, and safeguarding biodiversity.

With rising asset shortage, environmental change, and worry about ecological costs, the same old thing in the manner in which horticulture utilizes characteristic assets isn't a choice.

Making the cultivating frameworks of the rustic poor less powerless against environmental change is basic. Dealing with the associations among farming, characteristic asset preservation, and nature must be an essential piece of utilizing agribusiness for improvement.

Horticulture's commitments contrast in the three rustic universes: The manner in which farming works for improvement differs crosswise over nations relying upon how they depend on agribusiness as a wellspring of development and an instrument for neediness diminishment.

The commitment of horticulture to development and destitution diminishment can be seen by sorting nations as indicated by the offer of agribusiness in total development in the course of recent years, and the present offer of aggregate neediness in country territories, utilizing the $2-a-day destitution line.

This point of view produces three sorts of nations—three particular country universes (Table 3.1):

Table 3.1. Characteristics of Three Country Types, 2005

	Agriculture- based countries	Transforming countries	Urbanized countries
Rural population (millions), 2005	417	2,220	255
Share of population rural 1%), 2005	68	63	26
GDP per capita (2000 USS), 2005	379	1,068	3,489
Share of agriculture in GDP (%), 2005	29	13	6
Annual agricultural GDP growth, 1993-2005 1%)	4.0	2.9	22

Annual nonagricultural GDP growth, 1993-2005 (%)	3.5	7.0	2.7
Number of rural poor (millions), 2002	170	583	32
Rural poverty rate, 2002(%)	51	28	13

Source: Ravallion, Chen, and Sangraula 2007; World Bank 2006.
Note: Poverty line is $1.08 a day, in 1993 purchasing power parity dollars

- **Agriculture-based nations**—Agriculture is a noteworthy wellspring of development, representing 32% of GDP development by and large—fundamentally on the grounds that agribusiness is an extensive offer of GDP—and the vast majority of the poor are in provincial territories (70%).

  This gathering of nations has 417 million country occupants, predominantly in Sub-Saharan nations. Eighty-two percent of the provincial Sub-Saharan populace lives in farming -based nations.

- **Transforming nations**—Agriculture is not anymore a noteworthy wellspring of financial development, contributing by and large just 7% to GDP development, yet destitution remains overwhelmingly country (82% of all poor).

This gathering, encapsulated by China, India, Indonesia, Morocco, and Romania, has in excess of 2.2 billion country tenants. Ninety-eight percent of the country populace in South Asia, 96% in East Asia and the Pacific, and 92% in the Middle East and North Africa are in changing nations.

- **Urbanized nations**—Agriculture contributes specifically even less to monetary development, 5% overall, and neediness is urban. Overall, country regions still have 45% of poor people, and agribusiness and the sustenance business and administrations represent as much as 33% of GDP.

Incorporated into this gathering of 255 million country occupants are most nations in Latin America and the Caribbean and numerous in Europe and Central Asia. Eighty-eight percent of the country populaces in the two districts are in urbanized nations.

Nations take after transformative ways that can move them starting with one nation compose then onto the next. China and India moved from the agribusiness based to the changing gathering in the course of recent years, while Indonesia inclined toward the urbanized (Figure 3.3).

What's more, nations have sharp subnational land inconsistencies—for instance, many changing and urbanized nations have farming based locales, (for example, Bihar in India and Chiapas in Mexico) (Figure 3.3).

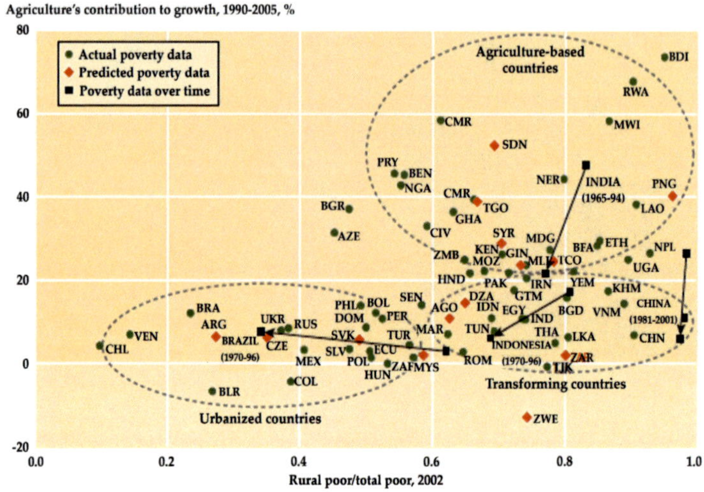

Figure 3.3. Agriculture's contribution to growth and the rural share in poverty distinguished three types of countries: agriculture-based, transforming and urbanized: Agriculture's contribution to growth, 1990–2005, in %.

Source: WDR 2008 Team Note: Arrows show paths for Brazil, China, India, and Indonesia

Grouping locales inside nations as indicated by their agrarian potential and access to business sectors demonstrates that 61% of the rustic populace in creating nations lives in favored territories—watered, sticky, and semi-humid regions with little dampness push, and with medium to great market get to (under five hours from a market town of at least 5,000).

In any case, 66% of the provincial populace in Sub-Saharan Africa lives in less-favored regions characterized as dry and semiarid or with poor market get to.

In five nations with itemized neediness maps, the destitution rate is higher in less supported regions, however, the vast majority of the poor live in favored zones.

So, utilizing horticulture to decrease destitution requires not just putting resources into less supported regions to battle extraordinary neediness, yet additionally focusing on the expansive number of poor in favored territories.

Heterogeneity characterizes the provincial world: Monetary and social heterogeneity is a characterizing normal for country regions. Huge business ranchers coincide with smallholders. This decent variety penetrates the little holder populace too. Business smallholders convey surpluses to nourishment markets and offer in the advantages of growing markets for the new agribusiness of high-esteem exercises. Yet, numerous others are in subsistence cultivating, mostly because of low resource gifts and troublesome settings.

Devouring a large portion of the nourishment they deliver, they partake in business sectors as purchasers of sustenance and as merchants of work.

Participation in these classifications is influenced by resource positions, as well as by sexual orientation, ethnicity, and societal position, as they suggest contrasting capacities to utilize similar resources and assets in reacting to circumstances.

Heterogeneity is found in the rustic work showcase where there are some low-aptitude, ineffectively compensated farming employments and few high-expertise occupations that offer specialists pathways out of destitution.

It is found in the country nonfarm economy where low-efficiency self- and wage-work coincides with work in unique endeavors. Furthermore, it is found in the results of relocation, which lifts a portion of the country poor out of destitution, however, takes others to urban ghettos and proceeded with neediness.

This inescapable heterogeneity in agribusiness and provincial society has profound ramifications for the open approach in utilizing horticulture for advancement. A specific strategy change is probably going to have gainers and washouts. Exchange progression that raises the cost of sustenance harms net purchasers (the biggest gathering of rustic poor in nations like Bolivia and Bangladesh) and advantages net dealers (the biggest gathering of the country poor in Cambodia and Vietnam).

Strategies must be separated by the status and setting of family units, assessing winning sexual orientation standards.

Separated approaches are composed not really to support one gathering over the other however to serve all family units more cost viable, fitting arrangements to their conditions and needs, especially to the poorest.

Adjusting consideration regarding the favored and less supported subsectors, locales, and families are one of the hardest approach situations confronting poor nations with extreme asset imperatives (Figure 3.4).

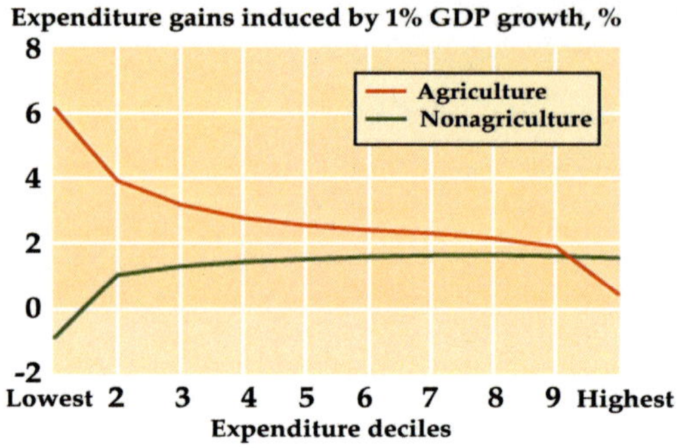

Figure 3.4. GDP growth originating in agriculture benefits the poorest half of the population substantially more.

Source: Ligon and Sadoulat 2007

Note: Based on data from 42 countries during the period 1981–2003. Gains are significantly different for the lower half of expenditure declines.

3.9. FARMER CASE STUDIES: SUCCESS STORIES SHOWCASE ENTREPRENEURSHIP AND INNOVATION OF GREENBELT FARMERS

The most recent research from the Friends of the Greenbelt Foundation centers around nine fruitful cultivating activities inside Ontario's almost two million sections of land of for all time ensured Greenbelt.

Home to 5,500 ranches, farmland makes up 43 for each penny of the Greenbelt. Agriculturists in the Greenbelt develop higher esteem crops than most Ontario ranches—with net homestead receipts per section of land in the Golden Horseshoe almost multiplying the Ontario normal ($1,756 contrasted with $939).

Some of the farmers' success stories:

Carncroft Farms: By working cooperatively and sharing some bits of knowledge and equipment, the Carnaghans are putting up more Ontario lamb for sale to the public: Luke and Jenny Carnaghan steered of Carncroft Farms, Luke's granddad's domesticated animals activities, in 2003. Today, lamb creation is the fundamental undertaking for this youthful entrepreneurial wedded couple living and cultivating in Ontario's Greenbelt.

Carncroft Farms, in Durham Region, has extended from its own crowd of 150 ewes to a noteworthy 700 ewe activity working together with two different accomplices. By sharing thoughts, equipment, and information, this creative three gathering joint effort has utilized its aggregate assurance and endeavors into a gainful plan of action.

Alongside the creative partnership, the Carnaghans moved their showcasing procedures from conventional domesticated animals sell off deals serving the Christian and Greek Orthodox Easter markets to all year lambing and promoting associations with five other Ontario sheep makers and a noteworthy Ontario sheep abattoir. Therefore, they can serve markets a year a year.

The Carnaghans credit the achievement of their generally youthful business to their quest of promoting training and business abilities, and the systems administration and tutoring of adjacent ranchers.

Receptiveness and fearlessness in the appropriation of various generation rehearse has likewise been useful. "We are on a decent way and set out toward proceeded with progress," the Carnaghans keep up.

Cooper's Farm and CSA: The Coopers have discovered their specialty in coordinate showcasing, and proceed to develop and differentiate their items and client encounter: Farming for over 20 years in Durham Region, Steve, and Lisa Cooper realized early on that diversifying was a crucial tool to be successful. They now run a Community Supported Agriculture (CSA) venture, take part in five farmers' markets, and offer an on-farm store and roadside stand.

Cultivating for more than 20 years in Durham Region, Steve, and Lisa Cooper acknowledged right off the bat that expanding was a critical apparatus to be fruitful. They presently run a Community Supported Agriculture (CSA) wander, partake in five ranchers' business sectors, and offer an on-cultivate store and roadside stand.

The Coopers cultivate 100 acres of land and lease another 300 sections of land close-by their activities in the for all time ensured Greenbelt. The grounds are utilized for their domesticated animal's creation and CSA, and in addition, field, feed, and developing some feed crops.

Their CSA was propelled in 2007 with 50 families and has now developed to 550 in the mid-year and 250 in the winter.

To diversify and grow their operations, the Coopers have also built an agri-entertainment/education business by setting up a corn maze and offering farm tours, as part of their goal to continue expanding their offerings. The Coopers consider their farm to be a work in progress, always changing and adapting to customer demands.

To broaden and develop their activities, the Coopers have likewise assembled an Agri-excitement/training business by setting up a corn labyrinth and offering ranch visits, as a component of their objective to keep extending their contributions.

The Coopers view their homestead as a work in advance, continually changing and adjusting to client requests.

These Greenbelt agriculturists were perceived with a noteworthy honor as Canada's Outstanding Young Farmers of 2010, a tribute to their advancement and ranch administration rehearses.

Meyers Fruit Farms: To remain beneficial, family natural product cultivate develops and enhances with the changing circumstances: Meyers Fruit Farms was begun by Fred Meyers' original migrant guardians in 1955 when they purchased farmland close Niagara-on-the-Lake. Today, Fred, and his sister Elly deal with the business, alongside Fred's child Jim and Elly's child Aron. A family-worked cultivate completely.

This Greenbelt cultivate incorporates 250 sections of land of claimed and 100 sections of land of rented arrive where the family develops scrumptious peaches, nectarines, pears, plums, crisp grapes, and wine grapes. To keep growing their activities, the Meyers have grasped enhancement and furthermore create nursery blossoms and herbs.

Around 70 for each penny of the homestead's flower deals are immediate to clients in the United States, with the rest of clients in the Canadian market. Their natural product is promoted crosswise over North America in association with a noteworthy neighboring organic product maker and distributor.

Officially supporting 25 full-time and 75 low maintenance occupations, Meyers Fruit Farms is in the arranging phases of a noteworthy development—a case of a business enterprise that helps make the Greenbelt a financial powerhouse.

Beverly Greenhouses – The VanderHouts are lessening their natural effect while developing their cucumber activities: The third era of a family-possessed task, siblings Jan and Dale VanderHout together run Beverly Greenhouses Limited, an effective greenhouse cucumber-creating ranch situated close Water down in Ontario is for all time ensured Greenbelt.

More than three decades back, the VanderHouts began focusing only on cucumbers, a standard and compelling practice in the present greenhouse vegetable industry in Ontario.

By practicing on cucumbers, the siblings have possessed the capacity to profit by just requiring one arrangement of hardware and not requiring a few varieties of bundling—both boosting their primary concern. Jan and Dale center on what they do well.

The province's greenhouse industry is eminent as a model as far as innovative headway and Beverly Greenhouses is in that spot with whatever is left of this industry.

New progressions in innovation utilizing cucumber fixing, bundling, and palletizing gear have restored a strong money related profit for their capital speculation. At a rate of 19–20,000 cucumbers in 60 minutes, their bundling line is exceptionally productive.

To guarantee their tasks have enough wellsprings of astounding water the siblings received a creative water reusing innovation for both natural and cost reasons.

With their eyes to the future and ensuring nature best of the psyche, they are focused on passing on their effective and imaginative activity to a fourth era best of the brain, "our goal isn't to leave here in 20 years and have everything debased – this is our sustenance," the Vanderhouts said.

Greenwood Mushroom Farms – Durham maker has mushroomed into one of the biggest undertakings of its kind in Canada: With a creative possession group of five individuals, this Greenbelt maker in Durham has mushroomed into one of the biggest endeavors of its kind in Canada—giving an amazing 360 employment.

In business for more than 50 years, Greenwood Mushroom Farms presently has two generation locales and a vast preparing and bundling activity. Getting the opportunity to be one of the nation's biggest makers wasn't without its difficulties and Greenwood is focused on finding creative arrangements.

The smell from their treating the soil activity has caused significant disturbance from neighbors. To address this issue, Greenwood is receiving inventive innovation to encase the activity.

Windmill Farms, an industry pioneer with a broad and modern conveyance system and conveyance framework, ships Greenwood mushrooms seven days seven days inside eastern Canada.

These shipments go to retail, nourishment administration, and pizza showcases and incorporate more than 100 neighborhood and particular Greenwood items.

Throughout the years, Greenwood has extended in a few stages progressively fusing automation, which has improved cost strength and create consistency.

Carron Farms Limited – A fourth era family cultivate drives development through re-marking and new item contributions to meet client intrigue: Jason Verkaik is the fourth era in his family to effectively cultivate in Ontario. In 1934, Jason's granddad and his two siblings moved to the Holland Marsh to clear land and start cultivating in the strikingly profitable claim to fame trim region which is currently secured by Ontario's Greenbelt.

More than 80 years after the fact, Carron Farm's vegetable generation presently covers a great 250 sections of land of Marsh and high land.

The Verkaiks put resources into a pressing line and a robotized pressing framework for onions, carrots, and beets. When Jason took responsibility for a family task, he began to re-mark the business, including another advertising approach in view of recounting his family's account of cultivating and where sustenance originates from.

Jason additionally spearheaded the Harvest Share Box program in his general vicinity. In just five years, Harvest Share has developed from 64 to more than 500 clients.

The Verkaiks give their very own portion vegetables in the week after week box, and source the rest from different cultivators of value delivery, or from the Ontario Food Terminal—all Ontario developed.

Not exclusively is Harvest Share self-maintaining, it has impelled Carron Farms into attempting new items, for example, Indian Red Carrots and multi-shaded beets, to meet the wants of customers, including new Canadians.

Preliminaries with different carrot assortments gave the information expected to present a multi-shaded carrot pack sold through Sobey's. In excess of 50 sections of land of East Indian Red Carrots and 100 sections of land of multi-hued treasure, carrots are presently developed in Ontario consistently.

Jason has situated Carron Farms for proceeding with development and he's focused on its prosperity: "There's a great deal of work still to do, and furthermore a ceaseless drive to continue onward."

Tigchelaar Berry Farms Inc. – Two siblings prevail by grasping development and reacting to shopper request: Tigchelaar Berry Farms Inc. embodies the way in which numerous effective homestead organizations have developed.

Dan and Jeff Tigchelaar are third era Ontario strawberry producers, at present based close Vineland in the Greenbelt. At the point when the customary strawberry business started to decay, they perceived the need to expand and enhance.

Subsequently, they began developing day unbiased strawberries with four sections of land of the yield in 1998. Since they aren't affected by day length, day neutrals can blossom and develop natural product consistently, as long as temperatures are reasonable. The measure of sunlight regularly controls the start of a strawberry plants blossoming process.

The Tigchelaar are pioneers in the berry segment, grasping advancement underway and always endeavoring to enhance yields and quality.

An innovative arrangement of coasting column covers advances pre-winter plant improvement and bloom bud commencement, along these lines supporting a late-winter reap.

What's more, to remain at the bleeding edge of generation innovation, the siblings direct on cultivating preliminaries and work on ventures working together with the Vineland Research and Innovation Centre.

With minor changes, the Tigchelaars' berry gear has been imaginatively adjusted to serve the necessities of a neighboring grape cultivator, along these lines creating another wellspring of income.

The local food development has created a customer ability to pay a higher cost for local food and the Tigchelaars are profiting by that over a more drawn out season with the day neutrals. "We can't contend on cost alone," says the Tigchelaars, "but since shoppers will pay more for a quality neighborhood item, we can survive."

3.10. CONCLUSIONS

Drawing together the exercises from 16 isolates look into thinks about is never a simple errand. The quickest perception concerns the heterogeneity that exists among cultivate based endeavors.

Decent variety can be seen among business people, who contrast with respect to their inspirations to begin new pursuits, the assets they control, the manner in which they take in, their desires and objectives, the kinds of items and administrations they create and the setting in which they work.

Independent of relevant varieties and their geographic and market situating, the improvement of homestead organizations requires generous entrepreneurial abilities.

These aptitudes can be found in the capacities of a few ranchers to animate their thought generative limit and to move themselves in investigating the potential uniqueness of their asset base. Albeit a few people may seem to have solid intrinsic abilities, the dominant part gains entrepreneurial aptitudes through training.

Closer associations with business sectors and clients, either eye to eye at agriculturists' business sectors and through collaboration with different performing artists in the esteem chain, or essentially through the use of ICT, gives chances to ranchers to build up an entrepreneurial introduction and additionally showcase introduction.

In spite of the continuous admonishments for agriculturists to wind up more entrepreneurial, a full comprehension of their sensible limits in such manner additionally requires a valuation for the limitations that ranchers may confront.

The family cultivates places confinements and additionally gives openings in the improvement of ranch-based endeavors. It gives the idea that new open doors might be even more frequently perceived and abused amid transitional periods in the family life cycle, instead of amid times of relative dependability.

While it is perceived that home-based business postures numerous difficulties for specialists, it likewise required a refined arrangement system to guarantee the fitting help at nearby, territorial, and between national levels.

Arrangements to help cultivate based business are especially testing given the heterogeneity both of the segment and of individual involvement in empowering and supporting business inside little and medium-sized undertakings (SMEs) in non-cultivate areas, this experience has not generally been moved into the cultivating parts.

One case of this lies in the part that arrangements may play in decreasing the hazard related with entrepreneurial wandering.

Inside the non-cultivate segments, the arrangement of start-up help as seed capital and business counsel and support has been broadly received, generally because of dynamic commitment and exchange between strategy creators and SME proprietor chiefs and their delegate association

However, the help for cultivating based endeavors seems all the more piecemeal – to a great extent as an outcome of the various and in some cases contending destinations obvious inside rural and provincial approach.

Entrepreneurial ventures inside the horticulture and ranger service parts frequently emerge because of network activities, proposing that an imperative approach measurement lies in supporting neighborhood entrepreneurial limit building.

REFERENCES

1. *Agriculture for Development*, World Development Report 2008, The World Bank.
2. Alsos, G. A., (2011). *The Handbook of Research on Entrepreneurship in Agriculture and Rural Development.* Edward Elgar Publishing Limited. Cheltenham, UK.
3. Nicoara, A., (2015). *Farmer Case Studies: Success Stories Showcase Entrepreneurship and Innovation of Greenbelt Farmers*. http://www.greenbelt.ca/farmer_case_studies.

CHAPTER 4

BEST PRACTICES IN AGRICULTURAL ENTREPRENEURSHIP

CONTENTS

Much of the sayings persist in the recent times regarding the much of the farmers becoming the businesspersons or the entrepreneurs or the occurrence of the entrepreneurship within the agriculture. Thus, there are two categories or the two parts of the entrepreneurship: the first one is known to be the possession of the skills within the management, which are required to get started so that they would be able to run up the business within the farm, which provides them with much of the benefits. In addition, the second one can be said to the possession of the spirit of the entrepreneurship, which is considered equally essential as that of the above one.

This chapter provides with the clear understanding and knowledge regarding the various best practices, which are being utilized within the agriculture by the various farmers who are mainly interested in becoming the entrepreneurs or the successful businesspersons.

By having the sound understanding of this concept, it is being hoped that the various workers who are being extended will be able to provide with much better support or the aid to the various farmers so that they would be able to possess the development or the growth of their own skills and also the spirit being required to be an entrepreneur.

4.1. INTRODUCTION

With the increment within the integration within the various markets and also the process of the globalization which is being present since very long time, the economy within the rural sector now possesses much more impact or the influence of the various farmers who are hoping to become the businessmen or the entrepreneurs.

These are particularly much responsive to the occurrence of the various within the demand of the market, which is being related to the various transformations or the changes occurring within the choices or the preferences, the fads, and the fashion within the society.

It has been put quite distinctively, as the various farmers who want to become the businesspersons or the entrepreneurs are being related to or are being connected with the various supply- chains, which are being integrated within the industry, and are being active for the purpose of the creation or the making up of the various novel networks.

These networks are being compared to other farmers are possessing much more tradition in comparison to that of these. It would not be such a big surprise or amazement that the studies or the research regarding various sources of creation or making up of the wealth or the money out of the agriculture, since the past twenty years and these are being related to the various policies as well.

They generally provide or present with the various strategies for the purpose of the growth or the development within the economy of the rural sector which all are being aimed or being targeted at the originating or the generation of the capacity within the enterprise, whereas it also tends to consider or acknowledge all the strategies which are being totally dependent on the capacity of the enterprise or the organization.

In accordance with the provided study or the researches, the farmers who want to become or are generally the entrepreneurs or the businesspersons, which are being shaped up by the provided strategies, are being much more receptive in relation to that of the various novel advancements or the technologies.

It also provides with the aid or the assistance from the various extensions, which all are being much more attentive towards the occurrence of the various transformations or the changes within the various prices and the accessibility or the availability of the credit as well as the version of the lesser number of the risks (Figure 4.1).

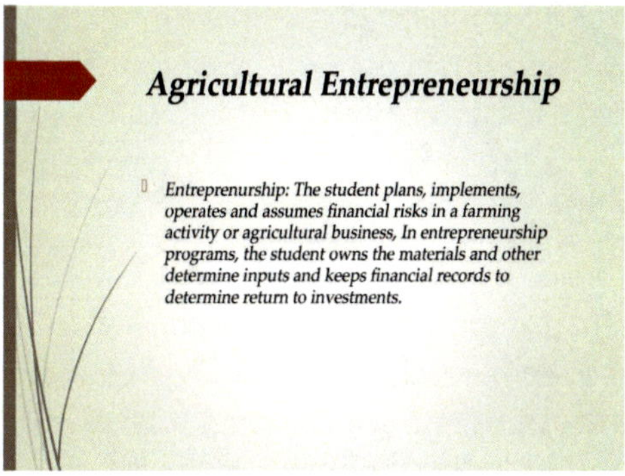

Figure 4.1. Agricultural entrepreneurship.

Source: https://www.slideshare.net/lerato25/entrepreneurship-powerpoint-slide

Within the various recent or the advanced economies which have been provided with the description above, are being known quite less regarding how the various innovations or the advancements buy the farmers, who are being the businessmen or the entrepreneurs, which all are being related to the pertaining partners of the various industries. It generally possesses the inclusion of the various sectors within the process of the processing up of the food, and the process of the research and the development or the growth within the economy of the various farmers who are being the entrepreneurs or the businesspersons.

The possession of one of the various views hold that being the farmer, who are being the entrepreneurs or the businessmen, generally tend to look out for much of the efficacy or the efficiency and also the various ways of doing the production so that they would be able to generate much of the profits or the gains.

They are also being open to the various advancements or the innovations and also the execution or the implementation of the innovation or the advancements, which are being in relation with their own partners.

It possesses the inclusion of the various farmers who are being the businesspersons or the entrepreneurs, who all are in search for the much of the wider or the greater interaction with the various sectors involving the process of the processing up of the food.

Therefore, all of these sectors being related to the economy are generally being marked up by the difference or the distinction, however, within the terminologies being related to the structure of the existence of the market.

Today, much of the farmers are being seen or are being looked up to as the takers of the prices whereas the various sectors within the processing up of the food, possess the behavior of being concentrated and also are not being much competitive.

Thus, it might become the case of the possession of the spurt within the drives of the various integration which all are having an increment of the various farmers, who are being the entrepreneurs or the businessmen, so that they would be able to get adapted or adjusted with the various advancements or the innovations within the technology within the sector of the processing.

In addition to it, the process of the interplay being persisting between the advancement or the innovation from both the sectors is also, basically, being influenced are getting affected by the various policies being made for the public.

These are being done with the help of the basic research being related to the process of the farming or within the sector of the agriculture, within the various institution so the various universities and the agencies within them being related to the process of the research.

Thus, it is generally accepted or agreed that the various researches being related with the agriculture of the sector of the public, by making up or creating the various basic knowledge of the science, having much growth or the development of the various new technologies.

Helping these technologies to be facilitated within the process of diffusion and helping in adopting the options or the varieties of the plant being modern and the various methods being involved in the process of the farming, have all attained or achieved the success, which is notable or remarkable.

Therefore, the literature sometimes provides with the view that all these pertaining benefits or the gains within the agriculture or the process of farming are not being completely reflected within the higher welfare being related with the social sector.

This happens as all these processes can be pessimistically or negatively being affected or impacted by the structure of the market and also might possess the dearth of the competition within the sector of the processing the downstream.

Therefore, though much of the knowledge or the information is being attained regarding the recent economy of the sector of the rural area, this still requires or needs to have much more to learn about or to gain knowledge about.

This particular paper generally tends to share or contribute within the recent literature, being in the theoretical format, and also is not being developed, within the specific area through the presentation and also the process of identifying and analyzing the model, being dynamic, of the various advancements or the innovations within the technology and also the process of the entrepreneurship within the agricultural sector.

This particular model, which is being dynamic, provides with or presents with the two sectors being related with the economy, which are: the sector which is not competitive and is related to the various industries, for example, the process of the processing up of the food, and also which is being considered to be the leader.

This also possesses the inclusion of the sector being related to the agriculture, which is being acknowledged or considered to be the one who is the follower.

Thus, this explicit or the particular closeness within the integration of both of the mentioned sectors, provides with the sign regarding the entrepreneurism by the various farmers who are being the businessmen or the entrepreneurs and also are being encouraged or motivated to have a spurt or an increment within the process of the production and also the various other margins.

Further, there occurs the process of the testing or the examination of the various pertaining determinants being related to the prices of the agriculture or the process of farming, the result or the output and the process of the research and the growth or the development within the sector of the processing up of the various foods.

Being provided with the various highlights regarding the current or the recent transformations of the changes being occurred within the structure of the various institutions of the research being related to the sector of the agriculture.

Also, the possession of the better or the sound understanding of the pertaining connection or the relationship between the various farmers who are being the entrepreneurs or the businesspersons and the process of the research and the development or the growth.

These are generally being acquired or attained by the process of the model which is being dynamic, and are being developed or enhanced within the society, is considered to be quite complex or critical.

The solutions, which are being provided, from the development of the formal model, within this particular research or the study, which helps in providing with the display of the various cycles, which are being quite explosive, for the process of the research and development or the process of the growth.

Wherein, it helps to possess a spurt or an increment within the sector of the processing of the research and development, which generally tend to develop or grow even wider or larger.

The solutions being generated from the model which is being formal and that had been created or being developed within the particular study, which provides with the display of the various cycles, which are explosive in nature for the process of the research and development.

It possesses an inclusion of the spurt within this particular domain or the sector of the expenditures incurred in the process of the research and development or the growth, which provides with the aid in growing, enhancing, or fostering even at a large extent. So as to ignore or avoid these various paths which are not that stable or uniform, the process of processing up of the firms generally tend to generate or provide with the control or the hold which is quite optimal with the various solutions or the ways which are quite much of stable or uniform

In addition, it happens only by having a setup of the various dynamic or adventurous paths or the ways of the process of the research and development or the growth and the main objective or the target it possesses.

Therefore, in addition to it, the various models which are being quite formal provides with the permit for doing the research or the study related to the of the various effect or the impacts of the races possessing patent, the research is totally based on the agriculture, the talent being present within the entrepreneur.

It also possesses an inclusion of the process of the pricing of the various final goods or the inventory on the process of the sector of processing the food within the sector of the research and development or the growth and also the various prices and the output regarding the sector of the agriculture.

All the aspects, which have been mentioned above, are of much utilization for the particular model in the process of the creation or the generation of the clear or the sound understanding or the information of the various connections or the relationships amongst the prices of the agricultural sector or the process of farming.

It also possesses the inclusion of the various sector of the process of the processing of the food within the sector of the research and development or the growth, the research that is being totally based on the agriculture sector, be it the public one or the private one.

In addition, the various prices being related to the various final goods or the inventory which are being provided with the help or the aid through the process of presenting the various data and also provides with the indication of the all optimistic or the positive connection or the relationship which are being predicted amongst the various variables be it the first or the one which is considered to be the last.

4.2. AGRICULTURAL ENTREPRENEURSHIP

There arises a question regarding the description or the meaning of the entrepreneurship within the agriculture or the process of farming. Therefore, in order to begin with or startup with it, there is no permanent or the fixed and description or the definition of definition regarding the business or the process of the entrepreneurship.

It also possesses the inclusion of the wide or a great variety of the meaning or the definition regarding this, can also be found or searched for.

In the regular or the daily language, the terminology of the word entrepreneur or the businesspersons is generally being utilized on an interchangeable basis with the various owners of the business or the enterprises, who does the startup, all those people who are generally being self- employed.

It also has an inclusion of all those people who are being the only trader or the sole trader, the various people who does the farming, which often tends to confuse or puzzle with the status, which is the position or the rank within the particular community or the society possessing the role or the part which is including the behavior or the nature being related to the specific or the explicit position.

The literature which is being totally dependent on the agricultural sector or the process of farming, within this particular aspect which are not much of the utilization or of help as it has been providing or presenting with the multitude of the various definitions with respect to the various operations of the entrepreneurs or the businessmen within the sector of the agriculture.

It is being represented or provided with the clear understanding of the attainment or the process of the acquiring of the quota for the larger milk or the acquiring or the attainment of the various greenhouses being in addition to it. These are particularly being up and the process of the running is not being taken into accordance or the consideration as the various opportunities in relation to that of the enterprises or the various businesses.

All the various things, which are being mentioned above, are considered the various arguments or the debates, which has to be put ahead or have to be done regarding the various pertaining reason for the definition or the clear understanding of the opportunities being the definition or the representation of the overarching.

Thus, it is being acknowledged or the being considered as the most attractive for the process of the entrepreneurship which has been included within the sector of the agriculture:

- It does not provide any kind of limitation or the restriction on the study or the various researches being conducted related to the process of the entrepreneurship within the sector of the agriculture to the various situations or the conditions, which are being specific or particular.

It possesses an inclusion of all the novel creation or the making up or the developing of the ventures. For example, much of the enterprises or the businesses being related to the sector of the agriculture, are already being within the persistence or the existence for many of the years or the decades.

- The process of the learning and also the growth or the development is being the heart and soul of the procedure of the entrepreneurship or the process of the business.

It is based on the various facts that generally all the people who are doing the farming, tend to exploit or spoil the various opportunities with respect to the entrepreneurship and various other are being due to the dearth within the various traits of the personalities of the various other people.

However, just because of this particular dearth in accordance with the particular competence and also the experience or the suffering.

- It thus possesses the recognition of the importance or the essentiality regarding the broader environment which is quite active or working which generally tend to keep all the businessmen or the entrepreneur engaged in this particular process.

The process of interpreting, having a sound understand and also the creativity, and the inclusion of the process being related to the core, within the process of the development or the enhancement within the opportunities.

Thus, all these are not being occurred or happening to be alone or in the process of isolation which are being affected or influenced by the wife of the people who does the farming, the various employees, all the competitors, the various networks and the partners within the chain or the services being provided for the process of extension.

Particularly, in the form of the addition, the main concentration or the focus is on the finding or the process of identification of the various opportunities and the pursuit, which is being related to it, are being acknowledged or being considered to as the core of the entrepreneurship within the sector of the agriculture.

It tends to put on much of the emphasis on the process of the creativity, the art of being the proactive and the aspect, which is totally being concerned with the networking of the activities being related to the various enterprises or the businesses.

It then helps in making the various researchers capable of shifting or moving away from the question which has been raised regarding who all the entrepreneurs are which is respect to the question being raised regarding what all work has been done by the various entrepreneurs. Therefore, it provides with the support or the aid so that it would help them in ignoring or avoiding the swamp being related to the various concepts of providing with the apt definition or the meaning of the true businesspersons or the entrepreneurs.

4.3. REASON FOR THE EXISTENCE OF AGRICULTURAL ENTREPRENEURSHIP

Earlier, the sector is related to the agriculture has been seen or looked up as the industry possessing the low technology with the dynamics which are restricted or limited, which are being dominated emphasized by the various number of the families whose size are actually small.

These are mainly put on with the concentration or the focus on pursuing or performing the various tasks instead of performing the various novel things or doing them.

Since the past ten years or the decade, this particular situation or the condition has been transformed or changed extremely because of the applicability of the process of the liberalization within the c=economy of the respective country.

These also have an inclusion of the protection which has been reduced or limited to protection or the security of the various pertaining market within the sector of the agriculture of the process of the farming, and also the changes or the transformation taking place on a rapid pace, which are being considered to be much more complex or critical, within the existing society or the community.

Thus, the companies being included in the sector of the agriculture or the process of the farming, there has been a rapid spurt within the economy that have to adapt to the various vagaries which are being related to the existing market.

It also includes the transformation or the changes occurring within the habits of the various consumers, the improved or the enhanced regulations within the environment, the various novel needs or the requirements regarding the quality of the particular product

The management of the supply of the chain, the security or the protection of the food, the process of the sustainability and also includes much more, are also being included within it.

4.4. OPEN-ENDED ISSUES

The important or the classic questions, which are being arisen within the various discussions or the debates regarding the occurrence of the entrepreneurship within the sector of the agriculture or the process of the farming.

The question is generally regarding the difference pertaining within the entrepreneurship within the sector of the agriculture of the process of farming from that of the process of the entrepreneurship within the firms, which are being related to the sectors, which does not possess any kind of agricultural activity.

The answer being generated can either be in the form of affirmation or the pessimism, which totally depends on the kind or the type of the question related to the research and the employment of the research related to the paradigm.

In the earlier days, the setup related to the style of working within the sector of the agriculture did not essentially provide with the behavior of the entrepreneurs or the businesspersons.

Since the past five decades so the fifty years, in much of the countries from the west, the sector related to the agriculture or the process of farming got a domain which was extremely or highly specialized on the basis of the efficacy or the efficiency or the productivity as well.

For example, in case of many of the countries, the occurrence of the modernization of the process after the war regarding the sector of the agriculture, was actually got much of success for the aims or the objectives it possessed, being quite real, so that they would be able to provide with the adequate protection or the security to the food.

4.5. UNDERSTANDING ENTREPRENEURSHIP IN FARMING

4.5.1. Farmers as Entrepreneurs

Sometimes, the question arises regarding the farmers being the businesspersons or the entrepreneurs and they are being doubted on this matter. However, yes, the various farmers, which are being present from the small-scale, all across the globe, provided with the view regarding the notable or the remarkable capability or the potential in order to possess the adaption.

All of these farmers tend to look up for the various better ways or the manner to provide with the organization of their respective farms.

They tend to provide with the trial regarding the novel crops and also the various cultivars, the animals being presenting or providing with the better condition, the various choices or the alternatives for the technologies so that they would be able to have a spurt within the level of the productivity.

Possessing the diversification within the process of the production, having a reduction or the minimizing of the pertaining risks and to have an increment within the benefits or the gains within this particular sector.

All these farmers have tended to become much oriented with the market and thus, have attained the learning so they could be able to reckon the risks being present so as to open up or develop the novel markets for the purpose of their own provided products or the services.

The farmers who are being the smallholders generally tend to do the process of farming for one of the provided four reasons (Figure 4.2):

- It is or explicitly for consuming it at their homes only, possessing quite rare chances of having a surplus within the process of the production.
- Much of it is being utilized or consumed at the home only, but with the possession of the intent regarding the selling up of the production, which is a surplus within the market.
- It is partially being consumed or utilized for the purpose of the market and partially for the purpose of having the whole consumption or the utilization at the home only for personal use.
- Otherwise, possessing it explicitly for the utilization or the consumption in the market only and not at home.

Figure 4.2. Ladder of intentions and reasons for farming.

Source: http://www.fao.org/uploads/media/5-EntrepreneurshipInternLores.pdf

In case of the first and the foremost rung of the particular ladder, it has been occupied by the various farmers who do the process of farming explicitly for the consumption within their particular homes only.

However, in case, there pertains a surplus at the time of the production, they would prefer to sell that out in the market, which actually happens quite rarely. Generally, most of the farmers are doing the struggle possessing the basic or much-required survival of themselves and their respective families, as they do not earn much to have a sound living.

They possess the dearth of the protection or the security and safety in terms of the water they are getting to drink, particularly their health issues, the food and the shelter, which has been provided to them.

It happens quite rare that they are in the state or the position to make up their mind regarding being the businessmen or the entrepreneurs as it requires capital or the investment and also the being the bodies of the performing the various tasks regarding the entrepreneurship within the business or the enterprise.

While, the farmers might become entrepreneurs from the soul or the spirit, but they certainly or generally tend to possess the dearth of the opportunities to do the process of farming being the businesspersons or the entrepreneur.

4.5.2. The Entrepreneurial Environment

Becoming a businessperson or the entrepreneur is being considered as the manner or the way in which the life is being led, and also the perception or the way of how people look at the world and what all thoughts they possess.

Generally, the businesspersons or the entrepreneurs tend to enjoy the advantage they are being provided with, of being quite independent and possessing the much- required freedom. These entrepreneurs tend to make up their own decisions about what all things are to be done and the manner in which they should be doing it all.

But the con would be counted as they have to face quite a lot of pertaining risks, they have to perform the various tasks remaining under the stress or the pressure and also are being held responsible for all kinds of the result, be it the good or the bad one, depending on the various decisions they made regarding this.

While all the people who are doing the farming and are being the entrepreneurs or the businesspersons are considered to be quite free as well as possess the required independence, thus, they staying alone are not doing all the work (Figure 4.3).

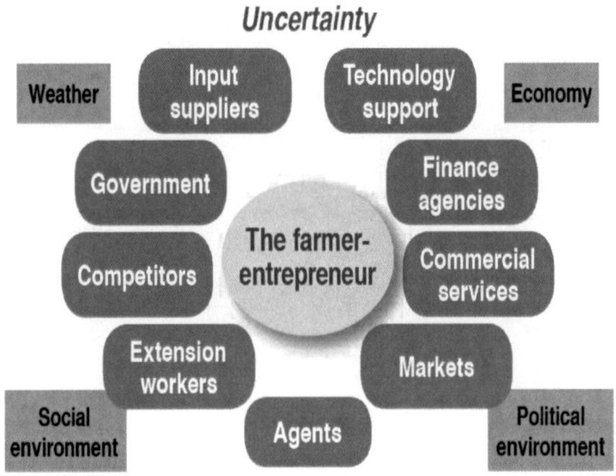

Figure 4.3. The world of the farmer-entrepreneur.

Source: http://www.fao.org/uploads/media/5-EntrepreneurshipInternLores.pdf

For all the farmers to tackle or cope up with the various pertaining risks, they also have to face various complexities as well as within the world where they generally tend to compete within the particular market, a proper development is being required by the various farmers who are being the entrepreneurs or the businessmen, the much needed or required spirit for the process of entrepreneurship.

The farmers possessing the spirit, which are being required for the process of the entrepreneurship, are quite energetic, full of enthusiasm, and often tends to take a various distinct decision after considering much of the care regarding the farm within the context or the value regarding the concept of the value.

It is in the context regarding the supply of the chain, which possesses a strong effect or the impact on the gains or the profits being made out of the various businesses being related to the process of the farming or the sector being related to the sector of the agriculture.

Therefore, all of these generally tend to occur or happen in the setting, which is considered the dynamic one, the mostly transforming or the changing environment and the possession of the uncertainties within the market across the various countries.

To make up with the assurance regarding the development within the business being related to the farm or the sector being related to the agriculture and possess the adaptation as per response in accordance with the pertaining changes or the transformations, the various people doing the farming and are being the entrepreneurs or the businesspersons are being required to:

- Maintaining with the apt determination or the concentration on the specific purpose being provided.
- Performing the work according to their bests so that they could be able to turn up all the events in accordance with their profit or the benefit.
- Making a proper utilization or seizing up of the opportunity and thus, making their best of it.
- Making of the entire system to perform the tasks accordingly so that they would be able to be provided with the favor.

4.5.3. Group Entrepreneurship

The process of the entrepreneurship can also happen or can occur in the areas wherein the whole particular group of the farmers generally tend to develop or establish a business or the organization being together.

All these farmers generally tend to possess similar or alike aims or the targets and the objectives to contribute or have a sharing of the benefits or the various pertaining risks as well.

Thus, the process of the ownership and possessing a control or hold over the whole of the organization or the enterprise, are generally being divided or segregated amongst the few members of the group.

In addition, this particular group is being acknowledged by the investor within the financial sector, the employee and the one who can take up the risks and bear them.

Thus, they tend to find or search for the protection or the security with the help of the various activities being related to the various groups which provide them with the permit regarding the process of the pooling up of the various resources, and also contributing or sharing up of the pertaining risks as well as the growth or the development of the safety net which is being socially utilized.

Therefore, in order to be quite successful, various enterprises being present in the groups should possess the similar skills, which are being required by the enterprises and the spirit, which are necessary to be present for being the individual businesspersons or the entrepreneurs.

Much of the members being present within that particular group tend to possess the want or the desire regarding being self- employed, having an encouragement or the motivating of taking up the various novel things to perform.

It also possesses the inclusion of the willingness within the people to take up with the already reckoned risks and the required mindset to always be present to have a look up at the various pertaining opportunities.

Thus, all of the people being present there much be willing enough so that they could be able to perform the tasks being together in the activity being related to the productivity which is quite common to all, and also to take up the various pertaining duties and also the responsibility for the generation of the result or the outcome.

4.5.4. Entrepreneurship Dynamics

The ideology or the vision being related to the process of entrepreneurship is said to be quite complex as well as critical. The skills which are being required by the various farmers much also be transforming or change from time to time and also enhance or develop so that they would be able to meet up with the demands of the process of management being related to that of the various businesses or the enterprises.

Thus, the process of the growth or the development within the enterprise being related with the process of the farming, as a business or the enterprise, generally tend to happen or occur in five stages (Figure 4.4):

- Having a proper establishment;
- Sustenance as well as having survival;
- Possessing an early development or the growth; and
- Having the maturity and possessing the probable decline or the reduction.

1. **Establishment**: The construction or the organization of the enterprise or the various businesses at this particular moment or the stage is generally being considered quite simple without possessing any kind of complexities.

The various challenges are being related to the various potential of the respective markets, the accessibility or the availability of the various pertaining resources and the much basic skills being in relation to that of the business.

Thus, all the farmers tend to need or require the various skills so that they could be able to do the negotiation with the various pertaining banks and the various other agencies so that they would be getting the much- required assistance or the aid to incorporate or establish their particular novel enterprise or the business.

2. **Survival:** Starting up or establishing a novel organization or the enterprise generally tend to provide with the display regarding the farmers possessing the various skills which are being required so as to become an entrepreneur.

   Thus, when they survive the first phase or the stage, it provides with an idea, which shows that the novel enterprise or the organizations possess a validity, which quite short- is termed.

   Thus, within the stage or survival, main concentration or the focus generally tend to be on the connection or the relation existing between the earning up of the income and the cost, which is being involved within this process.

3. **Early growth:** In case, the farmer takes up a decision regarding the novel enterprise or the business beyond the consideration of the survival, the various organizations or the enterprise being set has been developed or is being required to grow.

   So, in order to attain this particular level, the farer is supposed to grow up or develop the product possessing broader concept and also the base being related to the buyer, in the process of having an assurance regarding the business being quite profitable throughout, which is being related to the sector of the agriculture or the process of farming.

4. **Rapid growth:** In case the business or the enterprise being related to the process of farming, is working aptly being the very well-integrated business being related to the process of farm or being related to the sector of agriculture.

   Thus, it is being considered the stage or the position where they are quite capable of attaining or acquiring the much- required growth or the development being on the rapid p**ace.**

There is this one particular manner or the way to develop or grow is by having a spurt within the amount or the number of lands that have been planted and raising or increasing the livestock largely.

5. **Maturity (and possible decline):** Certainly, after all these steps, the business being related to the process of farming or the sector of the agriculture, will reach to the peak or the maturity.

This provides with the meaning that it happens that after such a point the business stops growing, developing, or even expanding.

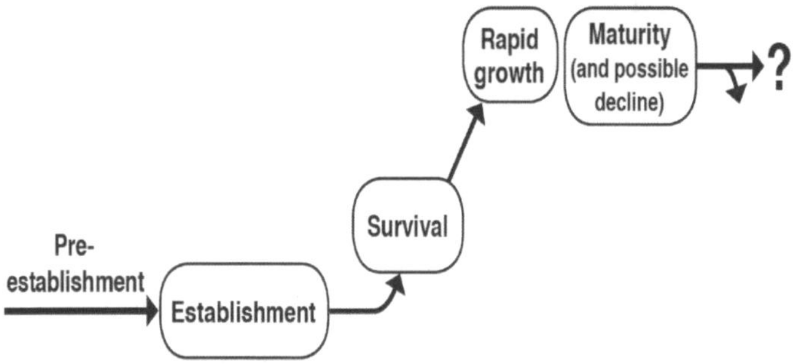

Figure 4.4. Five stages of farm enterprise development.

Source: http://www.fao.org/uploads/media/5-EntrepreneurshipInternLores.pdf

4.6. ENTREPRENEURIAL RESPONSES TO CHANGE

4.6.1. Capturing Value within the Value Chain

Most of the money in farming is made after farmers sell their product. Entrepreneurial farmers actively look for ways to capture the benefit within the value chain. This will increase profits. Farmer-entrepreneurs realize that capturing value requires producing for buyers and final consumers.

4.6.2. Enhancing and Managing Efficiencies in Production

Farmer-entrepreneurs continuously seek ways to improve efficiency, cut costs and increase productivity. It could be by reducing the level of money invested in equipment or looking for inputs that cost less, but give the same or better production results. It could be by changing the production system to one that is less costly.

4.6.3. Promoting Innovation and New Technologies

One of the ways to change production systems is by developing or adopting new technologies and practices. New technologies are being developed all the time. Government researchers, others by private agricultural companies, and others by farmers develop some. All these ideas and technologies need to be shared so that more farmers can benefit from them.

Farmers need to be encouraged to develop and adopt new technologies and share them with other farmers. Extension workers can facilitate partnerships among farmers, between farmers and researchers to work together to identify, develop, and test new technologies and practices to improve productivity and profitability.

4.7. EXTENSION SUPPORT FOR DEVELOPING ENTREPRENEURIAL CAPACITY

4.7.1. Training and Extension Support

Much of the specialized or explicit training is required or demanded so that they could be able to develop or grow the capacity of the enterprise amongst the various farmers or the people who do the process of farming.

So, in order provide a support to the process of learning by the method of doing it, the various requirements or the needs of the training to make or create much more opportunities for the people who do the farming.

This is being done so that they would be able to explore various novel ideologies, the various distinctive manner, or the ways of performing the tasks and to make the perspectives or the opinions much broader and wider.

4.7.2. Facilitating Entrepreneurship Training Programmes

Below are the various indicators provided which will provide with the aid or support in giving away the advice or the suggestion regarding the various programs for the process of training in the entrepreneurship:

- **Learning Outcomes:** The various outcomes of the process of learning are quite transparent or explicit and soundly defined and tends to match up with the level of the knowledge or information, the various attributes related to the skills and the behavior, which are being required at each stage in the process of the growth or the development.

- **Participants:** The various programs being conducted so as to provide the training, are being designed with a clear and also the sound understanding of who all people doing the farming will be interested in having a participation within it.
- **Selection:** It requires all the participants who all are being selected within this process, what all criteria have been utilized and many more. The main concern generally lies on the basis on which they are being selected:
  - Alike or the similar themes or the backgrounds and also the system of the process of farming.
  - Alikeness within the level of least skills and also the knowledge or the information being possessed by the various farmers.
  - Commonly providing with the requirements or the needs of the training.

4.7.3. Monitoring and Evaluating Performance

In case if it is being required to facilitate or present or provide with the workers being related to the process extension and also the training, will certainly be requiring the apt assessment of the various programs being conducted to provide the training in terms of the execution of that particular program.

It also possesses an inclusion of the satisfaction of the various people who are taking part in such programs and the learning that has been acquired or attained by the various participants.

Thus, it will certainly provide with the aid to assure that all the programs are being quite relevant or reasonable and provide with the apt information regarding the actual requirements and the concerns for the various farmers (Figure 4.5).

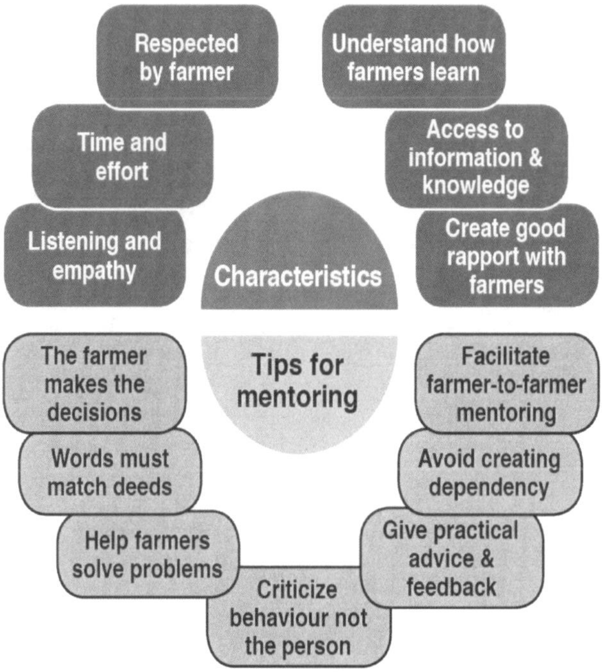

Figure 4.5. Characteristics of a good mentor and tips for mentoring.

Source: http://www.fao.org/uploads/media/5-EntrepreneurshipInternLores.pdf

4.8. TRANSFORMING AGRICULTURE THROUGH INNOVATIONS

Having an increment within the domain of the productivity and the sustainability of the various agriculture tends to depend widely on the process of having an engagement of the various people, who are quite young, within this particular sector or the domain.

Attracting much of the required attention towards the various advancements or the innovations and the energy.

Thus, this particular publication, which has been put all along by the various agencies within the country which are in collaboration with the various agencies and also provides with the information regarding the recognition and also what all have been achieved since then (Figure 4.6).

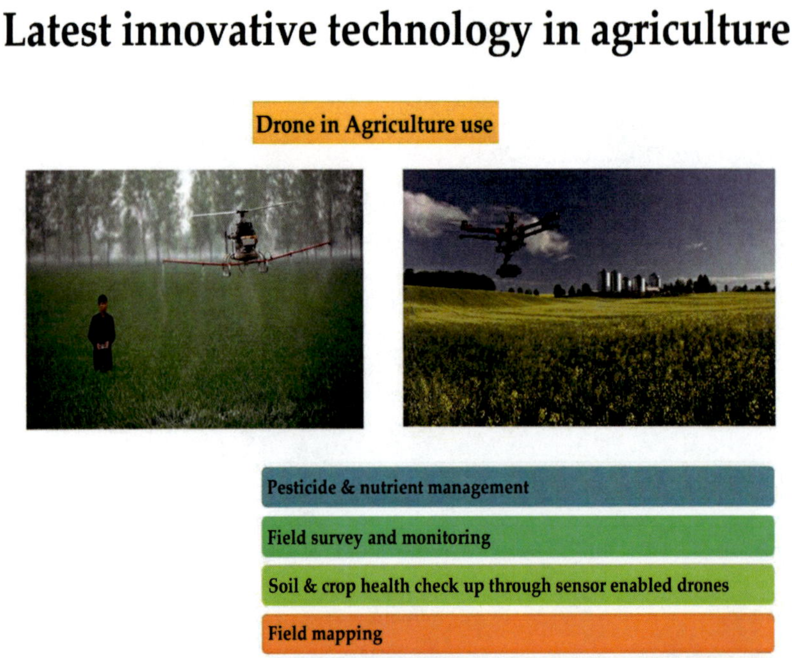

Figure 4.6. Transforming agriculture through innovative farm mechanization.

Source: https://slideplayer.com/slide/12327051/

4.8.1. Production

Agripreneurs use the high-speed internet in ICT to put information into the hands of those who most need it.

Considering the various countries across the globe, the various communities who are doing the farming tend to do a lot of struggle in order to attain or acquire an access to all the information, which have been provided which also, will help in fostering up the productivity and the sustainability as well.

For many of the farmers, it might be representing a description of not pertaining any knowledge regarding the recent or the latest advancements or the innovations in the field of the scientific research or the other advancements within the technology that would certainly help in having an increment of the yield with the least cost being added on it.

4.8.2. M-Shamba: Three-in-One

SMS connects farmers to agronomic information and local services.

It's been said that in the various parts of the world, especially the eastern one, possess extreme level of hunger and also at the similar time, much of the greater level of the occurrence of the harvests was being there in the various other parts of the world, most probably the one in the western parts, wherein the various crops are also being rotting.

Developing or growing up within the village of the farms, it has been seen that the gap in relation to the communication tends to pertain or exists. In addition, many of the people tend to grow a single specific crop for each year.

In addition, generally, it is the maize which has been mixed along with the various beans and that is the basic reason for creating up of the M-Shamba.

Therefore, in order to build up bridge the pertaining gap within the communication and also the information domain between the various institutions based on the research and also the various farmers who are being the smallholders so that they could be able to acquire knowledge regarding the growing up of the various crops in an efficient and also effective way (Figure 4.7).

Figure 4.7. M- Shamba bridges the information gap between research institutions and smallholder farmers.

Source: https://publications.cta.int/media/publications/downloads/1924_PDF.pdf

With all of the data collected, M- Shamba then shows the farmers how to search for local services and prices. The platform automatically sends this information to their mobile device.

4.9. CONCLUSION

The presence of the entrepreneurship within the sector of the agriculture or the process of framing certainly contributes or shares much of the features or the characteristics relating to the entrepreneurship of the generic domain.

However, it also possesses much of the distinctive features because of the various contexts, which are totally being specific of the sector of the agriculture or the process of farming.

Thus, a substantial or an important body of the literature regarding the entrepreneurship of the sector of agriculture or the process of farming has been grown or emerged and also the various discussions were pertaining regarding the various aspects of the entrepreneurship in the sector of the agriculture.

4.10. CASE STUDY

4.10.1. Introduction

Case study methods are frequently employed across a wide range of social science disciplines including, for example, farm management; business management, marketing, and psychology. Within these fields, they have long been regarded as both a legitimate and powerful way of exploring research and policy questions.

For many reasons, not least the imperatives and rewards of specialization, agricultural economists have. Generally favored econometric techniques dealing with a narrow range of information covering a large number of cases relating to the issue in question.

The hope is to derive some general conclusions about some aspect of a large number of cases, investigated in a narrowly disciplinary and relatively shallow way.

By completion, the case study method involves exploring fewer examples at greater depth. The aim of explanatory case studies, as distinct from and descriptive ones, is to investigate a small number of operating systems front many angles and in-depth to obtain insights into the likely impacts of the various obtained system to different but similar systems.

We would argue that once accepted as a legitimate research approach, case studies could be employed by agricultural economists to make a valuable contribution to research and policy development. The discounting of the case study method has generally taken two forms.

On the one hand, case study research has commonly been seen as an ~easy' option that may be useful for teaching purposes, but otherwise inferior to surveys because of limited ability to provide substantive insights beyond the particular case.

That is, there is a widely held prejudice against the apparent ability to generally beyond single or, at best from a limited array of cases.

On the other hand, the case studies that have been carried out with appropriate disciplinary breadth and depth are generally regarded as being complicated, messy, and all the rings are being considered, simply too hard to carry out and meaningfully use.

However, while there is a fair element of truth underpinning the latter viewpoint, it does not fully explain why the technique is largely ignored in agricultural economics yet widely used in other disciplines.

4.10.2. Issues in Choosing a Method for Projects

1. The policy problem: The extensive grasslands and woodlands of Australia have major degradation probability which is of concern from both a production and a conservation viewpoint (for example, to bill arid Gillies 1992, Department. of Environment, Sport and· Territories 1996).

The problems vary from region to region. While from, an economist' s viewpoint, there rightly be insufficient precision about the extent and location of the problems, the scientific quality and policymakers are in general agreement of a need to address this problem with ·some urgency.

To this and, some major programs are now in place in an attempt to address the problems {for example: Land and Water Resources R&D Corporation Remnant Native Vegetation Program, Environment Australia Biodiversity Group "Save the Hush" Program)

One issue or barrier to real progress, however, remains the belief (rationally grounded. or otherwise) that community conservation objectives are necessarily in conflict with Production objectives of both individual land managers, if not the community itself (e.g., Department of Environment Sport and Territories, 1996).

Despite the obvious resource management and: policy implications; the underlying reasons for this perceived conflict remain largely unsearched.

2. Some Important Consideration for Resource Management: In addressing land resource management problems. Issues of scale are important, from the viewpoint of both ecology and economics (e.g., HatringtQn, Wilson, and Young, 1984; Brown, and MacLeod, 1996).

Farmer decisions about either exploiting· or conserving total vegetation are typically made at the whole farm enterprise level. However; most agricultural R&D (including agricultural economics assessment of trial data) is conducted, at Smaller scales (e.g., plots land classes and occasionally paddock) and, thereby in order to have a look at the context within which such decisions are typically made.

Extension of the results related to it are non-surprisingly and typically provided at the same inappropriate technology transfer failure problems (MacLeod and Taylor, 1995) and also to fully analyze problems in land use decision-making, an obvious starting point is being considered within this particular case study.

R&D projects addressing land resource management issues now typically seek to include an extension component.

For example, all of the major industry funding agencies presently require a formal statement of how candidate projects will be structured to increase the likelihood of effective adoption of the results.

A problem remains, however, that conservation management R&D is notorious for generating failed expectations about effective motion on the ground by significant numbers of decision makers (for example, Pampel and Van Es, 1977; Vanclay and Lawrence, 1995; Macleod, 1997; MacLeod and Shulman, 1994; 1996; MacLeod, Shulman, and Taylor 1996).

4.10.3. Theoretical Issues and Research Preposition

Specifying the· theoretical issues necessarily begins with a. cleat outline of the specific issues of concern for the proposed study. Related to this process of considers issues to details is the basic need to clarify the underlying research questions, i.e. the 'who, where, what, why, and how questions addressed in a previous subsection.

It has already been suggested in case of study methods ideally address on how and why the questions rather than the other types and so this is actually a simple re-check on the basic appropriateness of adopting the case study approach.

From this exposition of the issues and the instances of the efficacy of the appropriate, populations can be developed which generally tend to plan the situation. Rival propositions may also be developed.

In this stage to be explored within a study, the format propositions and the various forms of the data. Collection and the final, which is being related to the various forms of the results.

4.10.4. Tests of Validity and Reliability

For the various particular research designs which are being included under this case study or otherwise there are essentially four· basic facts of logic that might be applied to assess its quality. These are:

- Construct validity: of the appropriate definitions and operational measures for the theoretical propositions being studied;
- Internal validity: appropriateness for establishing the credible causal relationships;
- External validity: generally, convincingly specifying the domain to which the findings can be generalized; and
- Reliability: the ability to repeat the findings if the same methods are being supplied.

REFERENCES

1. Crosthwaite, J., MacLeod, N., & Maicolrn, B., (1997). *Case Studies: 'Theory and Practice in Agricultural Economics*. [eBook]. p. 21. Available at: https://ageconsearch.umn.edu/bitstream/135749/1/fiche005-report047.pdf [Accessed 16 Aug. 2018].
2. Hailu, M (2016). *Innovate for Agriculture*. [eBook]. p. 100. Available at: https://publications.cta.int/media/publications/downloads/1924_PDF.pdf [Accessed 16 Aug. 2018].
3. Kahan, D., (2012). *Entrepreneurship in Farming*. [eBook]. p. 1. Available at: http://www.fao.org/uploads/media/5-EntrepreneurshipInternLores.pdf [Accessed 16 Aug. 2018].
4. Lans, T., (2013). *Agricultural Entrepreneurship*. [online] Available at: https://www.researchgate.net/publication/260985365_Agricultural_Entrepreneurship [Accessed 16 Aug. 2018].
5. *Promoting Agricultural Entrepreneurship among Rural Youth*, (2014). 2nd edn. [eBook], p. 5. Available at: https://www.iyfnet.org/sites/default/files/library/GPYE_RuralEntrepreneurship.pdf [Accessed 16 Aug. 2018].
6. Ricardo, F. J., & Mixon, G. Jr. F., (2016). *Farmer-Entrepreneurs, Agricultural Innovation, and Explosive Research and Development Cycles*. [eBook]. p. 11. Available at: http://file:///C:/Users/acer/Downloads/admsci–06–00013%20 (1).pdf [Accessed 16 Aug. 2018].
7. Sancho, F., (2010). *Agricultural and Rural Entrepreneurship*. [eBook], p. 15. Available at: http://repiica.iica.int/docs/b2032i/b2032i.pdf [Accessed 16 Aug. 2018].

CHAPTER 5

BUILDING ENTREPRENEURSHIP SKILLS

CONTENTS

The analysis of entrepreneurship has grown effectively and now is becoming the first concern in educational research, both for the volume of researches undertaken and the courses taught. While building up its own boundaries, this area of research has confided on other disciplines for much of its content. Economics, business education, and cognitive psychology have been effective in informing the effective methods in entrepreneurship research. Academic research in this kind of field has provided efficient knowledge to analyze the causes of entrepreneurial tendency (Gartner, 1985; Brockhaus and Horwitz, 1986; Bygrave, 1989; Baumol, 1993; Young and Sexton, 1997; Timmons, 1999; Rae, 2000).

Theories mainly based on severe quantitative research unpredictably prove the nature of an entrepreneur, the entrepreneurial activity and its impact on economic development. Still, there is a pressing requirement for a qualitative area of research to establish theories of how and why entrepreneurial activity happens. For example, whether entrepreneurial resolute is primarily decided in advance by steady personality traits or it is possible to promote entrepreneurship inclination towards self-employment through a sensible education program.

Developing skills and abilities among the youth population is offering a serious matter for educators and curriculum developers. Ironically, the country having an advantage of being a young nation, has a broad majority of students drop out at various levels of school education and at the same time are basically lack of employable skills and competencies. The chapter is concerned with the list of skills required for an entrepreneur and the impacts of them on entrepreneurship.

5.1. INTRODUCTION

Nowadays, entrepreneurship can be described as a major part of a strategy to expand the economy. It can be commenced as a challenging and demanding task, either internally within the organizations, either externally, by the production of new businesses, which are sustainable in a market and in a complicated economic environment.

Examples included are studies on the impact of entrepreneurship in transition economies. Therefore, it is examined that entrepreneurship is linked with two different aspects:

- Production and development of new business or innovative firms that is entrepreneurship; and

- Development of an activity in an innovative approach that is intrapreneurship. In the context of this chapter, the focus will be on the first part, correlating entrepreneurship to the production of innovative businesses.

The main aim of the research was to construct a model of skills development for entrepreneurs, which will be presented and discussed in this chapter. The chapter briefly explains the concept of entrepreneurship, followed by the presentation of the methodology used for constructing the skills development model, and the analysis and discussion of the results obtained from various researches.

5.1.1. Concept of Entrepreneurship

The source of the word 'entrepreneurship' is grounded in the Latin language, from the words pretenders and apprehension. Entrepreneurship can be assumed as an individual or collective system and internal or external to the structure of the organization, developing something new, from an understanding of ideas to the development of a business.

The approach of applied entrepreneurship argues that an entrepreneurial firm aims at innovation, is open to risk and aggressive in relation to its competitors. Entrepreneurship has evolved as a concept, and in the level of importance placed on the sustainability and development of the economy.

Various experts think that the entrepreneurs are the people who are eager to take risks which other people are not. Others describe them as people who start and construct successful businesses of their own.

Analyzing these definitions, entrepreneurship doesn't certainly involve starting your own business. Many individuals who don't work for themselves are identified as entrepreneurs within their organizations.

Becoming a successful entrepreneur is not an easy task. So, how does an individual successfully take benefit of an opportunity, while another, having the same level of skills and knowledge does not? Do entrepreneurs have a distinct genetic makeup? Or do they function from a different vantage point that somehow directs their decisions for them?

While many researchers have researched the subject but there are no definitive answers. What we do know is that the successful entrepreneurs seem to have certain characteristics in common.

These characteristics are further divided into four categories:

- Personal characteristics;

- Interpersonal skills;
- Critical and creative thinking skills; and
- Practical skills.

5.1.2. Entrepreneurial Capacity

Being an entrepreneur and the development of an entrepreneurial culture goes ahead the fear of risk and the stain of failure, which influence absolutely the context of entrepreneurship.

Enforcing ideas is not an easy method even though we can say that the profile of an entrepreneur(s)is critical to define a business idea and execute it successfully, but it is important to analyze that there is also the possibility of developing entrepreneurial traits and characteristics and here come the educational institutions that must play the main role, very early in the development of knowledge and skills correlated to entrepreneurship.

The example of an innovation is a practice, which can come into teaching from the earliest years of school, because it is a significant tool of entrepreneurs, utilized to examine new opportunities for business or a different service or product.

It is very important to analyze the method used to reduce the risk, explore for new sources of innovation, utilize creativity tools, and acquire information from the market; these are the skills that every entrepreneur or potential entrepreneurs require to have.

Every day we observe the event of the birth and death of companies, services, products, and processes, and the aim of entrepreneurship learning is to explore and to consistently seek new business or new practices, which add value to the market and simplify the economy.

In this perception, entrepreneurship is constructed based on the different kinds of skills broadly studied in the literature and are referred to as soft and hard skills. Entrepreneurship skills are the focus of analysis and the basis for constructing the skills development model.

The soft skills can be described as the behavioral skills needed for the application of hard skills and knowledge in organizations. It also suggests that soft skills are a group of skills and talents of an individual. Other authors classify the soft skills such as:

- Cognitive skills;
- Social and personal skills; and
- Interpersonal skills.

With concerned to soft skills essential to managing entrepreneurial projects, correlates that are skills and practices of successful managers. He states "the emphasis of the future have to be in leadership skills and interpersonal management practices that ensure project success." To recognize the soft skills of the entrepreneur, a qualitative methodology is used generally.

Analysis of the research done: It is expected that the profile of the entrepreneur enfolds the belief that entrepreneurial characteristics can be learned or acquired throughout the life. Based on various studies, ranges of characteristics that are associated with the entrepreneur profile are presented:

- Capability to attain goals, mainly planning and autonomy to achieve results.
- Self-confidence to attain defined goals and defeat obstacles.
- The sense of responsibility for the success or failure of the business.
- Capability to learn to utilize failure as a learning experience in a positive manner.
- Capability to develop and maintain a successful business through effort and hard work.
- Capability to develop innovative new methods to the market, new products and services, and new concepts.
- The capacity of initiative recognizing and implementing new business opportunities.
- The aspiration to achieve goals.
- Endurance to face hurdles and reset strategies.
- Capability to establish a range of contacts, to develop and maintain business relationships.
- Social and relational skills to demonstrate interpersonal relationships.

5.2. DEFINING ENTREPRENEURIAL SKILLS

Entrepreneurship is defined as the ability of an individual to turn innovative ideas into action. It involves innovation, creativity, and risk-taking, as well as the capability to plan and manage projects to achieve the goals and objectives. It is seen as crucial to economic growth, competitiveness, and promoting innovation.

Promoting entrepreneurial attitude supports the development of new firms and business growth. Moreover, entrepreneurship skills do provide benefits regardless of whether an individual sees their future as starting a business.

They can be utilized across an individual's personal and working lives as they include 'tenacity, creativity, initiative, teamwork, a sense of responsibility and understanding of risk.' What comprises entrepreneurship skills has been the matter of much discussion.

Entrepreneurial skills are not correlated to a specific discipline, occupation, or qualification, unlike other important economic skills.

However, the major attention on entrepreneurship education and enhancing entrepreneurial skills has brought more agreement and analysis of entrepreneurial competencies and abilities. The OECD has recognized three main groups of skills required by entrepreneurs:

- **Technical skills:** Environment monitoring, communication, problem-solving, technology implementation and use, interpersonal, organizational skills.
- **Business management skills:** Decision making, planning, and goal setting, human resources management, finance, marketing, accounting, quality control, customer relations, negotiation, business launch, compliance with regulations skills, growth management.
- **Personal entrepreneurial skills:** Innovation, risk management, leadership, self-control, and discipline, persistence, change management, strategic thinking, and network building.

This mixture of the skills attributes and creative workers and commercial managers require competencies diversely. Additionally, entrepreneurs need knowledge of the fields in which they function (i.e. an IT, construction or catering entrepreneur will require knowledge of those specific sectors or occupations) (Figure 5.1).

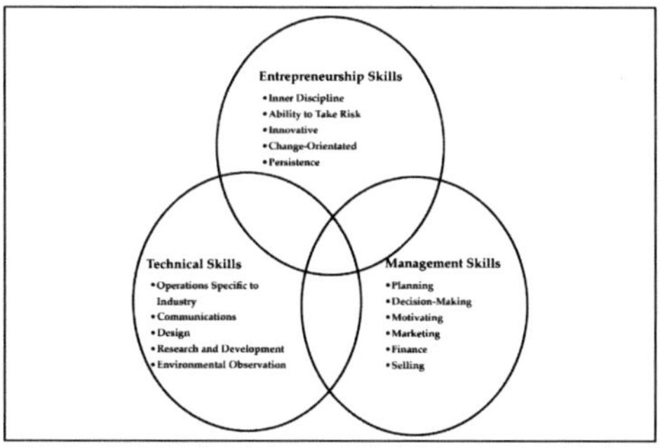

Figure 5.1. Entrepreneurial skills.

Source: https://www.semanticscholar.org/paper/Enhancing-Entrepreneurial-Skills-with-the-and-of-an-Maseko-Dehinbo/6ea8cf3f800c2ae22a9b0b061779bef729ee6364

5.2.1. Who Acquires Entrepreneurship Skills?

While a distinct group, entrepreneurs usually tend to be male, aged 40–55, educated to tertiary level and examine the idea of becoming self-employed as possible. Eurostat data for 2013 signifies that less than one-third (31%) of self-employed people in the European Union were female.

However, country-level aim differs essentially. In Malta, Ireland, and Romania less than one-quarter of the self-employed were female. In Latvia, Luxembourg, Lithuania, and Poland, at least 39% of self-employed people were female.

Some commentators analyze these differences as resulting from gender-specific hurdles that limit the choice of becoming self-employed rather than obstacles to starting a business per se (for example, capital, business idea, risk, and skills).

Having an idea of business and approach to finance are the most important elements when deciding to start a business but so are having an appropriate business partner and role model.

5.3. DEVELOPING ENTREPRENEURSHIP SKILLS

Around twice as many individuals who regard self-employment as a possible prospect in the next five years had followed an entrepreneurship course (34% compared to 18%). Still, less than half of the population feel that their school education aided them to better understand the function of entrepreneurs in society (47%) or gave them the skills and know-how to authorize them to run a business (41%).

Just 28% of Eurobarometer respondents acknowledged that their school education made them curious to become an entrepreneur; although this estimate ranged from 65% in Portugal to 17% in Germany and the United Kingdom.

Inducing entrepreneurship skills includes developing:

- Particular technical business-running skills, for example, developing and executing a business plan, budgeting, and accounting.
- Strategic or transverse skills linked with entrepreneurship. An ability to make decisions based on balanced risk analysis and information analysis, identifying and executing opportunities for the growth of business; following market developments and managing the services and products offer.

The European Commission identifies entrepreneurship as a major ability for life. The 'Rethinking Education' communication states that every young individual should benefit from entrepreneurship education, involving 'at least one practical entrepreneurial experience before leaving mandatory education.'

Entrepreneurship is a key feature of the Europe 2020 strategy. Therefore, Member States have been invited to stimulate the development of a coordinated method to entrepreneurship education all over the education and training system.

A various number of countries has produced entrepreneurship education strategies and there is an evolving body of knowledge about good and effective methods in learning and teaching. Inspirational teachers, institutions, and ministries are working harder to harmonize learning and the procurement of transversal entrepreneurship skills, behaviors, and knowledge.

Research has recognized scope for further development in the sector of education in relation to analyzing entrepreneurship, education, and working methods and the analysis of the outcomes. The European Commission is continually working together with teachers, experts, and the educators to identify good practice and enable its wider distribution.

5.3.1. Effective Ways to Build Entrepreneurial Skills

Take a different path: Creativity is defined as the ability to assume things differently and to develop solutions where there are gaps. To construct your creativity skills, purposely try something new, something which others won't do. Try to read unusual books.

"Creativity is the root of entrepreneurship."—Karndee Leopairote, Thammasat University. Watch a movie of a different language. Travel to an unusual spot. Talk to people who are out of your circle of comfort.

"The Big Short" is a movie that shows how several investors and opportunist entrepreneurs managed to get benefits from the 2008 financial crisis by going against popular opinion.

Start a business: There is nothing like real-world experience. Whether you run a firm on the side or full-time, you bring the opportunity to enhance your skills, for example, business planning, sales, and marketing and negotiation.

"You don't learn to walk by following rules. You learn by doing, and by falling over."- Richard Branson.

Stick with challenges: Every successful entrepreneur has educated to produce their tenacity muscles and perseverance. The life of an entrepreneur is never smooth glide, and it requires guts to keep going when people doubt your capabilities.

"It's not that I'm so smart, it's just that I stay with problems longer."—Albert Einstein. To construct diligence, develop a goal or challenge which is meaningful and do not give yourself to the quit. Alternately, give yourself a deadline to intent towards. For example, if you want to develop a better blog, make a commitment to write 1,000 words every day for a year.

Delay gratification: The founder of Tesla, Solar City and SpaceX, Elon Musk, fix himself a challenge to live on only $1 a day to observe if he had what it took to lead a life as an entrepreneur.

Entrepreneurs have to be ready for numerous failures and about zero rewards before they finally hit the jackpot. To train yourself to be capable to delay satisfaction, start small. Say "no" to the extra donut. Use your old car instead of going into debt to buy a new one. On the weekends, wake up at 5 a.m. And work on your business ideas instead of staying in bed.

Manage your own finances: Understanding elementary finance is important in running your own company. You don't need to be an accountant, but you must at least be capable to understand the basics around cash flow, profit, and loss, and assets.

Come forward to lead: The potential to lead a team and stay coordinated is important when you become an entrepreneur. You can begin by searching for volunteer and leadership opportunities around you. Volunteer to lead a Meetup group, begin a fundraising project for your favorite non-profit organization or get absorbed with your local community board. Alternately, coach a local children's sports team or just plan your mother's birthday party.

Practice communication skills: The best entrepreneurs have well-informed about how to communicate their dreams and passion in an alluring way, both online and offline. To learn how to speak publicly, accompany a Toastmasters group, offer to speak at workplace parties or even host at your friend's wedding.

To enhance your online communication skills, be active on your social media accounts, blog, create a newsletter on your favorite topic or set up an online Facebook group. The more frequently you put yourself out there, the faster your communication skills will grow.

Learn from a mentor: The advantage of a mentor is precious when it comes to constructing your entrepreneurial skills. Rather than making all the mistakes by yourself, why not learn from someone else who has already made them?

Mentors can be seemed as great sounding boards for your thoughts and ideas and they can also be fantastic cheerleaders when the going gets challenging. If you are lucky, you may get a mentor willing to prepare you for free because they believe in you and want to give back. Some mentors will be happy to train you in exchange for you helping them out in their own business. Some may offer a paid service.

If you are at the early stage of your career and young enough then try to request for internships as those are major opportunities to wet your feet in the real business world.

Work in sales: Sales play a crucial role in success, sustainability, and survival of a business. You can get the best product in the world but if you don't know how to sell it, it is useless. "To me, job titles don't matter. Everyone is in sales. It's the only way we stay in business."—Harvey Mackay.

One of the simplest ways to learn how to sell is to have a sales role. It really doesn't an issue if you are trading Tupperware at parties or selling complicated technology solutions, you will select valuable skills, such as learning how to ask questions of your prospect and researching your target market.

Get involved with different entrepreneurs: Whether it's attending entrepreneurial conferences, events, meetups or seminars; spending time with various entrepreneurs will help you to grow in your own entrepreneurial skills.

If you have the guts and time, you can challenge in an entrepreneur competition like Shark Tank where any voids in your entrepreneurial skills will be informed very quickly. Such a demanding experience will also integrate your speed of learning.

Help others with their businesses: Being an entrepreneur is all about resolving problems with the resources that you have. The more you help others to solve problems with their own businesses, the more your own skills will grow. "You will get all you want in life if you help enough other people get what they want."—Zig Ziglar.

Such as, you can develop an accountability group for entrepreneurs where you both can help each other out. In this process, you learn and grow together.

Keep on learning: "I am still learning."—Michaelangelo. Keep your learning and development activities. There are so many courses available online, both free and paid, that demonstrate a variety of entrepreneurial skills such as Skill Incubator, Udemy, and Udacity.

With such a simple approach to resources and knowledge, there really is no excuse that you can't construct skills to become successful as an entrepreneur (Figure 5.2).

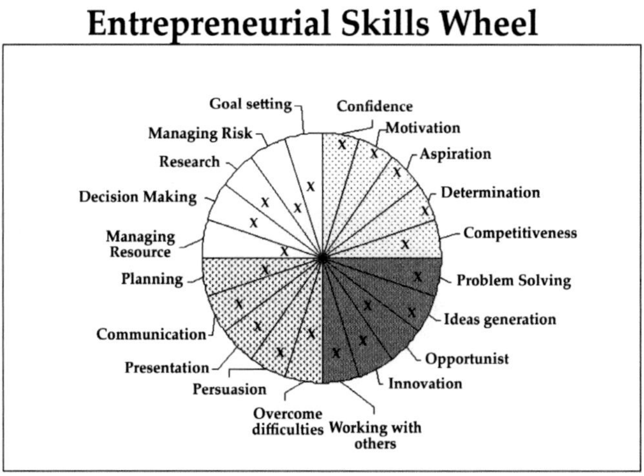

Figure 5.2. Entrepreneurial skills wheel.

Source: https://slideplayer.com/slide/801060/

5.4. CHARACTERISTICS OF SUCCESSFUL ENTREPRENEURS

5.4.1. Personal Characteristics

Examine personal values, beliefs, and characteristics, and analyze what do you have the mindset of becoming a successful entrepreneur. Here are few personal characteristics defined:

- **Optimism:** Be an optimistic thinker. Optimism is really an advantage, and it will help you to get through the difficult times which many entrepreneurs feel as they find a business model that works for them.
- **Vision:** Try to observe the points where the things can be further improved. Grasp the wide picture and explain this to other individuals. Try to produce a fascinating vision of the future and then encourage other individuals to involve with that vision.
- **Initiative:** Have an eagerness to do something and inherently start problem-solving or business encouragement projects.
- **The desire for control:** Try to enjoy being in charge and making decisions. Be self-motivated and effectively lead others.
- **Drive and persistence:** Be energetic and self-encouraged to work hard for an even longer time and hence realize your aims and objectives.
- **Risk tolerance:** Be able to take risks and make effective decisions when the facts are quite uncertain.
- **Resilience:** Be resilient. This will definitely help you to adapt yourself when the situation doesn't go as planned earlier. Try to grow and learn from failures and mistakes. Do not avoid the actions to escape yourself from failure. Try to overcome the fear of failure.

5.4.2. Interpersonal Skills

Being a successful entrepreneur will require you to work closely with people. This is the case where it is crucial to be able to make healthy relationships with the customers, team, suppliers, investors, shareholders etc. Few individuals are more gifted in this field than others, but, fortunately, you can learn and enhance these skills. The types of interpersonal skills you'll require include:

- **Motivation and leadership:** Be a leader and motivate other individuals to deliver their vision. Let other individuals follow you. Be able to delegate work to others. As a successful entrepreneur, you'll require to depend on others to get beyond a very early stage in your business – there's just too much to do all on your own!
- **Communication skills:** Be able to do all types of communications. You required to be capable to communicate well to market your vision of the future to team members, potential clients, investors etc.
- **Listening:** The ability to listen carefully is required in entrepreneurship. If you are able to listen, then you can become a good entrepreneur. Make sure that you have the skill of empathetic listening and active listening.
- **Personal relations:** Become emotionally intelligent. More you are emotionally intelligent, more it will be easy for you to interact and work with other individuals. Well, the emotional intelligence can be improved.
- **Negotiation:** Be a good negotiator. You do not only require negotiating sharp prices, you also required to be able to resolve differences between people in a positive and mutually beneficial way.
- **Ethics:** Do you deal with people based on respect, integrity, fairness, and truthfulness? Try to lead people ethically or socially. It will be very hard to construct a happy and committed team if you deal in a shabby way with the staff, suppliers, and customers.

5.4.3. Critical and Creative Thinking Skills

As an entrepreneur, it also requires coming up with new and fresh ideas and make effective decisions about opportunities and potential projects.

Many people think that you're either born creative or you're not. However, creativity is a skill that you can develop if you invest the time and effort.

- **Creative thinking:** observe and analyze the things from a wide variety of perspectives and then come up with original and fresh ideas. Various creativity tools are there which can help in doing this.

- **Problem-solving:** Be good at problem-solving skills. Always come up with alternate solutions to the current problems. You need to be familiar with the tools like Cause and Effect Analysis, CATWOE, the 5 Whys Technique.
- **Recognizing opportunities:** Always recognize coming opportunities. Plan to take benefit of the opportunities recognized.

5.4.4. Practical Skills

The practical skills and knowledge are required to produce goods or services efficiently and run a company (Figure 5.3).

- **Setting of the goal:** Try to set goals regularly. Create several plans to achieve those goals and then carry out the plan.
- **Planning and organizing:** Induce skills, abilities, and talents which are required to achieve the goals. Coordinate with different people to execute the plan and achieve the aim efficiently and effectively. Try to develop a well thought and coherent business plan, involving learning and developing from accurate financial forecasts.
- **Decision making:** Be good at decision making. Make the decisions based on appropriate information and by analyzing the potential consequences. Be confident while making decisions. Grid analysis, Six Thinking Hats and Decision Tree Analysis are the core decision-making tools.
- You require knowledge in different areas when starting or running a business. For instance:
- **Knowledge of the business:** Have a good general knowledge of the main functional fields of a business that is sales, finance, marketing, and operations, and are you capable to manage and operate others in these fields with a reasonable level of competence.
- **Entrepreneurial knowledge:** Gain the knowledge about how entrepreneurs increase the capital. Understand the abrupt amount of hard work and experimentation, which may be required to find out the business model which works for you.
- **Opportunity-specific knowledge:** Try to understand the market before entering into it. Bring your services and products to the market.

- **Venture-specific knowledge:** An entrepreneur requires to know what things should be done to make the business successful. The specifics of the business must be known by the entrepreneur.

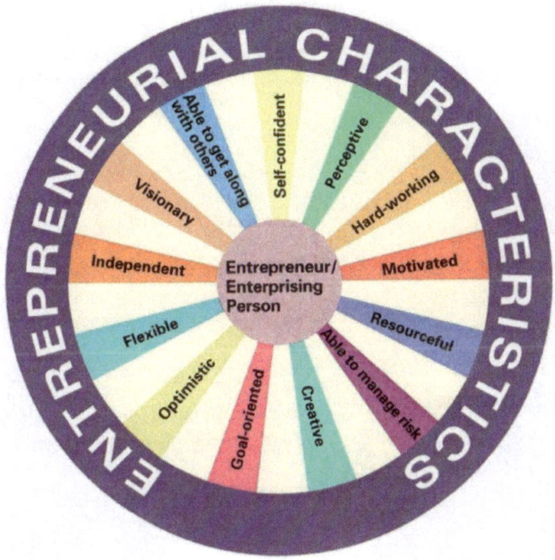

Figure 5.3. Characteristics of a successful entrepreneur.

Source: https://thesocialmediamonthly.com/characteristics-successful-entrepreneurs/

5.5. THE SKILLS THAT MAKE ENTREPRENEURS EXTRAORDINARY

Amy Wilkinson, the famous author of the book, "The Creator's Code: The Few Essential Skills of Extraordinary Entrepreneurs" once attended a birthday party in New York City and she found some prominent entrepreneurs who were doing really well in their respective field. Some of those entrepreneurs were; founders of Google, eBay, and Gilt Groupe.

One of these entrepreneurs from the White House Trade Office at the time made Wilkinson think about various things related to entrepreneurship. The main question that crossed her mind was what had made these entrepreneurs successful when so many others had failed.

Wilkinson in her book stated, "I decided to study the most successful entrepreneurs to figure out what skills they had in common," Thus, from that point of time she began her research odyssey: 200 interviews with leading entrepreneurs, including the founders of LinkedIn, Spanx, Chipotle, and Airbnb.

Almost 10,000 pages of interview transcripts, more than 5,000 pieces of archival and documentary evidence, and some 4,000 pages of academic research were evaluated and deeply analyzed by Wilkinson and she noted down few skills shared by the most successful entrepreneurs and wrote a book about them.

According to her, "extraordinary entrepreneurs are creators, who aren't necessarily born with an ability to build and scale companies successfully." The hard work of such successful entrepreneurs is dependent upon these skills:

5.5.1. Find the Gap

For any successful entrepreneur and to have an on-going business it is important to first understand the gaps that have not been filled with talent and skills. For example, Steve Ellis desired of establishing a Mexican fast-food restaurant that was, at the same time, the antithesis of fast food: made-to-order with high-quality ingredients.

In founding Chipotle in 1993, he also created a new dining category, which had the facility of fast casual. Ellis found the right kind of opportunity and analyzed those skills and things, which went impossibly for various people when they thought for the same idea of business. This is how he found the gap and started to dig an innovative way to fill the gap. This is how an innovative and successful entrepreneur works.

5.5.2. Drive for Daylight

To become a successful entrepreneur, it is important to have a clear vision and the person should focus upon the future of the business he is going to start-up with. This sounds similar to a racecar driver focus and the way he keeps his eyes fixed on the road ahead while seeking opportunities to pass a competitor.

Successful entrepreneurs are least bothered about the shortcomings of their track or lane and are least bothered about their competitors. Their focus is clear, and the vision is fixed. For example, the founder of Chobani yogurt, Hamdi Ulukaya, raised his company from zero to $1 billion in annual revenue in five years by making decisions with an eye toward the future. That included buying and rehabbing a decaying yogurt plant despite having only a few thousand dollars of available cash. Today, Chobani Greek yogurt sales are about half of all yogurt sales nationwide.

5.5.3. Fly the OODA Loop

Col. John Boyd, a U.S. Air Force fighter pilot, during the Korean War came up with a new idea of the framework to have rapid decision-making capabilities in fast-changing environments. According to him, the acronym for "observe, orient, decide, and act" was the right kind of framework for people who come across fast-changing environments.

For entrepreneurs, Wilkinson says, "this means continually updating their assumptions and moving quickly from one decision and iteration to the next." David Sacks, PayPal's first COO, once had a one-on-one conversation with Wilkinson, and he stated that "PayPal's team would look at new features that competitors were building and quickly iterate."

When Dotbank.com, for example, gave a $10 bonus to those who signed friends up for its service, PayPal rolled out a similar offer within a week, and added a $10 bonus for the new customer, too.

5.5.4. Fail Wisely

According to Wilkinson, "There's a lot of hero worship in startup circles, but they've all failed. Every single person I met with talked about failure." This illustrates that the person who succeeds in becoming a well-established and respected entrepreneur knows that he or she has the flexibility to avoid catastrophic mistakes, this could only be done when there is a series of smaller mistakes, and they handle these smaller mistakes intelligently.

Such entrepreneurs test ideas in low-risk environments as an inexpensive way of gathering insights to determine whether a product or idea will take off. According to Wilkinson, "Some concepts work, and some don't, but in either case, the results make creators smarter and more resilient."

5.5.5. Network Minds

To attain successful entrepreneurship, it is important that the person striving to handle an Agri-business should have capabilities to solve multifaceted problems by taking help from people coming from diversified culture and different level of perspectives.

This would help in bringing several ideas together on one platform to overcome the issues that arise in growing their businesses.

In any business, whether it is of agricultural background or non-agricultural background, there are various solutions to a problem as a problem could have different facets, but one solution is all it takes to overcome the hurdle in a business line.

The solution could be of any form, it could be of passing an opportunity on to a colleague, it could be writing few extra words for a friend, or preparing a strong proposal for a new project, all these solutions would further take a form of a reward for the business and its achievements.

According to Wilkinson, "entrepreneurs who build positive relationships by looking for ways to provide help gain a competitive advantage." "This was a surprise to me," she explains. "It's the result of the rapid transmission of information, especially through social networks, so our reputations are 100% known. That forces better behavior."

Wilkinson once stated that she was in a state of shock when she realized that the most successful entrepreneurs are still not comfortable with the achievements they have had in the past few years and they crave for striving.

The most successful entrepreneurs tune out naysayers. "These people are building new things and solving new problems all the time," she says. "They have had to learn to get comfortable with being uncomfortable."

5.6. DEVELOPMENT OF ENTREPRENEURIAL SKILLS

5.6.1. Model Development Skills

Key characteristics required for building a model of personal skills have been identified as skills, attitude, and creative thinking.

An entrepreneur should have cognitive skills, social, and relational skills, technical, and management skills which are very crucial in creating and running a business.

- Creative thinking, formulation, and problem-solving, decision-making, and initiative together constitute Cognitive Skills.
- A person's communication skill, capacity to work as a team and the adaptability to new situations comes under Social and Relational Skills.
- Technical skills include concepts of technical expertise like working methods, processes, and contexts.

- Knowledge of potential strategies adopted for the company, capacity planning and setting strategic goals and ability to make strategic decisions on company policies according to the various areas of expertise come under Management Skills.

Apart from the above-mentioned skills, an entrepreneur must have the knowledge of business development, which comprises of the strategies adopted by the company, products and/or services the company will be dealing into, corporate image management like human resource management, financial management, marketing etc. and the decision taken upon what should be the formal and informal structure of the company.

Skill development is needed to understand the mechanisms, which are crucially required for the creation and implementation of technological, organizational, and market innovations. It is necessary to keep the knowledge of specifics of market internationally through various modalities, export, licensing, franchising, joint venture, and foreign direct investment.

Analyzing the market environment, competition, and the company's internals, the best marketing strategies should be identified for business development.

Proper care and decision should be taken in creating a corporate brand name (such as advertisements, media publicity images etc.) for the products and or services the company is dealing with. There should always be some scopes to innovations in the products and or services provided by the company as per the requirements of the market needs.

Company's management systems such as business management, marketing, human resource, management, financial management, risk management etc. should be properly designed to give it a well-defined structure. It also enables the smooth functioning of the system (Figure 5.4).

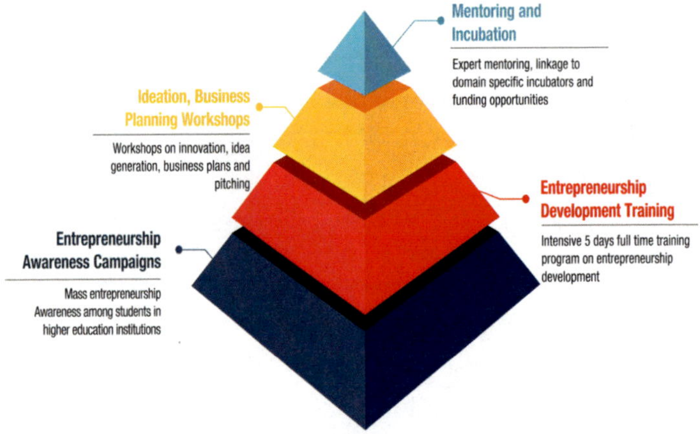

Figure 5.4. Entrepreneurship development framework.

Source: http://ka.ictacademy.in/Entrepreneurship-development.aspx

5.6.2. Implications for Practice

The contribution of entrepreneurship to the economic growth of the country: As far, the developing economy is concerned, where there are very few scopes of market growth, an entrepreneur helps largely.

It creates businesses, increases the transportation services which ultimately boosts up the economy of the country, promotes export and import of products and services creating scope for internationalization of firms, brings globalization resulting into foreign investments, creates competition in the market which further gives rise to price modification and ultimately the country's economy gets help in being boosted up.

However, entrepreneurship is not an easy task to be done. There are many companies, which establish, grow up their business, and are shut down due to some functional failures and other factors.

Roles and responsibilities in creating an entrepreneurial culture: In several companies, it is found that there is a lack of entrepreneurial culture in terms of company's policies, which neglect the respect of their employees, customers, and other people concerned, they do not have properly framed ideas and good business practices.

There are several factors are responsible to create the entrepreneurial cultures and develop the entrepreneurial skills:

- Government Policies- Several government economic,

financial, educational policies are responsible for promoting entrepreneurship.

- Schools and Universities- They play a vital role in developing entrepreneurial skills in the people through their education systems.
- Companies- The higher management is responsible for the implementation and management of organizational culture among the employees, which promotes creativity and innovations.

The government, schools/universities, and businesses- Collaborating with these entities encourage and promotes a new attitude towards risk.

5.7. ENTREPRENEURSHIP AWARENESS BUILDING

Basically, entrepreneurship is about being an idealized self-starter leadership, one who is capable to analyze opportunities to give solutions in the form of process, service, product or technology to fulfill the requirement and thereby collect and employ all important resources including finance, manpower, infrastructure, technology, and others to construct a successful enterprise in thc desired field of business.

An entrepreneur has a vision of making benefits and also has a greater vision of creating employment, helping people in creating their potential, devoting to the society and at the same time assuring the growth of his or her organization and the connected people.

At the primary stage, one might have an impulse to become an entrepreneur. But generally, the students will not have an idea of what it means to be an entrepreneur. Therefore, various awareness programs introducing the idea and the concept of entrepreneurship in schools and colleges as well as through focused training programs help the youth make up their minds to chase entrepreneurship as an option to a career.

5.7.1. Entrepreneurial Building

Being an entrepreneur demands for performing a multi-disciplinary and multi-functional role in managing the enterprise. A successful entrepreneur is capable of identifying business opportunities and is energetic in his approach in the development of the business opportunities into the successful business enterprise.

This demands for thinking out of the box and extraordinary creativity, leadership, foresight, ability to think ahead, ability to gather resources, ability to create options, manage resources to attain the goals as well as negotiate and resolve the issues both internal to the business as well as with external environment. Being energetic is an attitude of the mind.

Entrepreneurship is all about being capable to attract the best talent, to contribute leadership and guidance to grow and feed the talent, manage, and harness the talent pool too.

By the way of training one who has a basic ambition towards entrepreneurship can be carving to become a successful entrepreneur. Capability to think creatively, looking at the micro and macro environment, as well as options, transacting with alternatives and directing processes on all fronts without losing out on the concept, can be sharpened by providing Entrepreneurial training (Figure 5.5).

Figure 5.5. Skills of the entrepreneurial building.

Source: https://www.slideshare.net/krishnashah5891004/ram-power-point-presentation

5.7.2. Personal Empowerment Skills

Development and personal empowerment are important for every individual and most significantly for an entrepreneur. These skills apply to the ability of an individual to perform in his or her work area. One requires to have the significant job function or technical skill to follow a particular vocation but along with the functional knowledge and skill, it is the personal permission of the individual that helps him or her succeed in the enterprise.

The training of 'Personal Empower' skills affects mind training to the entrepreneur, helping him to learn and focus, to concentrate, to analyze and be objective in his thinking. Positive thinking along with learning to be affirmative and assertive as well as being equalized in thinking at all times, being quietly active, ability to control stress and work under pressure are some of the skills which are the must for every successful entrepreneur.

An entrepreneur can also be defined as a leader. By providing training, he can enhance leadership skills as well as groom oneself and construct a positive image as well as enhance the self-esteem. Personal development in the fields of time management, self-discipline, being self-motivated and enthusiast are very important for a young entrepreneur who is making his incursion into business. All these skills can be learned through specific training.

A good leader is confident and at the same time capable to encourage and encourage the others too. He has got to have major patience & wisdom, listening skills, emotional intelligence and more essentially a Win-Win attitude towards work and life.

An entrepreneur who has the ability to think big for oneself and for all of his team and take the whole organization to attain the heights and recognize their potential is a successful individual too. For being a good visionary entrepreneur, training is required which helps to sharpen his abilities, construct additional skills as well as make him bring out his true potential too.

5.7.3. Entrepreneurship Skills Required to Overcome Barriers to Growth

This is still a major topic of debate that whether the entrepreneurs are born or made. However, it is generally accepted that there are natural 'born' entrepreneurs, but there are also a few researchers who believe that entrepreneurship is a skill, which can be learned.

Drucker (1985) debated that entrepreneurship is a process and that "most of what you hear about entrepreneurship is all wrong. It is not magic; it is not mysterious, and it has nothing to do with genes. It's a kind of field of study and, like any other discipline, it can also be learned."

If a person agrees with Drucker's concept of entrepreneurship, then it understands that education and training can play a major role in its development. In a conventional understanding, entrepreneurship was strongly linked to the creation of a business and hence it was argued that the skills needed to achieve this outcome could be enhanced through training.

Currently, entrepreneurship is being considered as a way of behaving and thinking that is appropriate to all parts of society and the economy. Such an understanding of entrepreneurship now wants a different approach to training.

The educational process required in today's world is one which helps to cultivate an individual's mindset, skills, behavior, capabilities, and can be enforced to create value in a series of contexts and environments from the charities, public sector, universities, and social enterprises to associate new venture start-ups and organizations.

Lichtenstein and Lyons (2001) debated that it is essential for service providers to identify that entrepreneurs come to entrepreneurship with various levels of skills and therefore each entrepreneur needs a different 'game plan' for developing his or her skills.

In addition, they advised that the development of skills is a qualitative change, which appeals to some level of transformation on the part of the entrepreneur.

Actually, it has been debated that developing these skill-sets will induce enterprising persons who must be equipped to accomplish their potential and build their own futures, whether or not as entrepreneurs (NESTA, 2008).

Kelley contested that any educational training must empower people not just to prosper skills to start a business but rather to be able of behaving entrepreneurially in whatever role they take in life. This approach is quite wide, but it catches the critical philosophy of exclusive entrepreneurship education and training programs needed if countries are to produce an increasing pool of people who are interested to behave entrepreneurially.

5.7.4. Youth Entrepreneurship Building Skills

Constructing a society which is responsive to, enhancing, and favorable to Youth Entrepreneurship is the liability of every society as well as the Government. Motivating the youth to become entrepreneurs has many advantages in terms of resolving issues and problems related to unemployment in the economy as well as paving way for the growth and innovation of the country too.

Considering the requirement of creating and maturing youth entrepreneurship, many countries have taken up strategic actions and executed plans focused on imparting entrepreneurial skills and knowledge through education.

Without any doubt, such a move provides the thrust and the freedom for the youth to aim and attain their dreams. Leadership cannot be defined as a born talent but can be captured through training and learning and too.

Absolute entrepreneurial studies and training should involve subjects like Entrepreneurship Awareness Building Skills, Business Planning, Business Management, Entrepreneurial & Personal Empowerment skills as well as Personal Empowerment skills.

The ambition to grow a business is not an aim for all entrepreneurs and therefore those who do review their future in this way should be allowed bespoke support to make sure that they have the best chances of succeeding.

5.8. CHALLENGES AND FUTURE DEVELOPMENTS

Various attempts have been made over the past two decades to conclude the characteristics, which define high growth firms. Also, defines that how these characteristics can be duplicated amongst a wide range of entrepreneurs.

Storey's (1994) conclusion that a small number of high-growth firms produced real employment growth has focused the minds of researchers and policy-makers the same way to seek to analyze the key entrepreneurial skills that required to be developed by growth-orientated entrepreneurs.

Following the comprehensive review of the literature undertaken, the following are the principal entrepreneurial skills, which required to be developed amongst growth-orientated entrepreneurs:

- **Customer-Orientated** – Entrepreneurs should be dedicated to creating customer value through the arrangement of innovative services or products.
- **Strategic development** – Entrepreneurs must learn how to select the best strategy from a number of market strategies that can authorize their chances of success for e.g., Choice of Market, Constantly Innovate, Customer Driven, Differentiation/Focus, Exporting, Highest Quality etc.
- **Financial management** – Entrepreneurs should learn the skills needed to access supplementary venture capital for example how to structure a proposal.

- **Human resource management** – Entrepreneurs required appreciating and understanding the need to improve the HR practices of the firm and to provide financial incentives to employees.

The development of such skills is not enough in themselves as the following conditions must be there if any progress is to be attained:

- The entrepreneur should be highly motivated to grow the business.
- Peer-to-peer mentoring from successful entrepreneurs is a crucial element of any training process. Mentors should also be motivated to perform this role.
- Entrepreneurs should be provided with enhanced access to networks, international markets, and finance.

Probably, the major finding to come from the review of existing program is the function of mentoring and how the mentors should be people who have already attained success within the area identified by the growth-orientated entrepreneur.

This person has the access to networks and an experience, which enables to the growth-orientated entrepreneur to enhance their horizons. The mentor also acts as a role model and strengthens the belief of what ambitions can be achieved.

5.9. CONCLUSIONS

The main aim of this study was to recognize the skills required for developing a culture of entrepreneurship. The review of concepts of entrepreneurship, which is based on various methodologies, allowed analyzing personal and professional skills.

Personal skills include social and relational skills, cognitive skills, management, and technical skills whereas professional skills include strategy, which the company will adopt, services, and products, the corporate image, the management systems: financial management, human resource management, marketing, among others and the formal and informal structure of the company.

With an account of the future studies, it is essential to acknowledge that entrepreneurs require taking risks, having the capability to overcome issue and difficulties, take the action and reach objectives. The growth of entrepreneurial skills is crucial to develop a culture where entrepreneurship is something natural becoming a complex part of our evolution and a new model of the economy.

5.10. CASE STUDY: INSPIRING THE INNER ENTREPRENEUR IN STUDENTS: A CASE STUDY OF ENTREPRENEURSHIP STUDIES IN TUAS

5.10.1. Abstract

This part introduces a case study of the entrepreneurship studies in the Faculty of Telecommunication and e-Business at Turku University of Applied Sciences. There are three levels of entrepreneurship studies in our faculty:

- is the description of entrepreneurship?–to raise the awareness and motivation towards entrepreneurship;
- Student co-operatives and Support centers for training the students with the skills to manage and run a business; and
- Become an entrepreneur.

We describe existing recommendations for higher education to organize entrepreneurship and reflect our solutions to these recommendations.

5.10.2. Introduction

Entrepreneurship is increasingly recognized as an important driver of economic growth, productivity, innovation, and employment. The importance of entrepreneurship to Europe and Finland has been identified in many reports and studies published since 2000.

The European Union's Lisbon Strategy stated for example that new firms and SMEs are the major sources of growth and new jobs. Furthermore, the European Commission's Green Paper "Entrepreneurship in Europe" in 2003 stated that education should contribute to encouraging entrepreneurship, by fostering the right mindset, awareness of career opportunities as an entrepreneur and skills.

Entrepreneurship education should focus upon developing understanding, and capacity for pursuit, of entrepreneurial behaviors, skills, and attributes in widely different contexts. Nevertheless, it demands considerable changes in pedagogy, modes of assessment and the appropriate training of staff.

5.10.3. Entrepreneurship Education Recommendations for Higher Education

Europe's competitiveness, innovation, and economic growth depend on being able to produce future leaders with the skills and attitudes to be entrepreneurial in their professional lives, whether by creating their own companies or innovating in larger organizations.

Entrepreneurship education is the first and arguably the most important step for embedding an innovative culture in Europe. At the same time, the European Commission's report declares that currently the teaching of entrepreneurship is not yet sufficiently integrated into higher education institutions' curricula.

Clearly, for entrepreneurship to embrace the 21st century, educators must become more competent in the use of academic technology, and expand their pedagogies to include new and innovative approaches to the teaching of entrepreneurship.

5.10.4. Entrepreneurship Education in Engineering Education

It is no surprise that majority of entrepreneurship courses are offered in business schools. However, it is questionable whether Business Schools are the most appropriate place to teach entrepreneurship: innovative and viable business ideas are more likely to arise from technical, scientific, and creative studies.

Entrepreneurship needs to be expanded across the campus – especially to the technology and science departments, where many innovative ideas and companies originate. Technical and scientific universities are potential sources of start-ups and spin-offs. Increasingly, business and technical faculties are linking efforts to encourage the exchange of skills and ideas among students.

The challenge is to build inter-disciplinary approaches, making entrepreneurship education accessible to all students, creating teams for the development and exploitation of business ideas, mixing students from economic and business studies with students from other faculties and with different backgrounds.

5.10.5. Entrepreneurial and Entrepreneurship Education in Our Faculty

The strategy of Turku University of Applied Sciences defines that entrepreneurship and applied research and development will be connected to teaching. The University tries to actively inspire the inner entrepreneur and increase number of enterprises.

The entrepreneurship is characterized as expert entrepreneurship. Furthermore, the strategy defines that business studies will be increased in all disciplines. Our faculty's model for entrepreneurship education has three phases:

- What is entrepreneurship? Practice Enterprise + Real life projects;
- Learn real entrepreneurship in a safe environment: Student cooperatives + Education Support Center; and
- Become an entrepreneur: Personal Mentors + Student competitions.

Besides these special entrepreneurship education possibilities, the curricula have been designed in a manner that entrepreneurial competencies are strengthened. Actually, implementing CDIO has supported these very much.

Implementing design-build project in the curriculum makes student to reflect their skills and competencies. These projects are quite close to the projects in real enterprises. They train students to work on real projects and make them understand certain business rules.

The increased use of active learning methods together with integrated learning experiences improve student's problem-solving skills, communication, and negotiation skills – all important entrepreneurial competencies.

5.10.6. Discussion

Higher education institutes were recommended to integrate the teaching of entrepreneurship in their curricula. As we described earlier we have integrated many different ways to learn entrepreneurship in our curricula.

In addition, the ways to learn entrepreneurship are not very traditional rather fresh and inspiring solutions. The paths suggested by the Ministry of Education are partly followed. The first path that focused on intentions is strongly followed: the basic idea of the Practice Enterprise – the course is to strengthen the intentions towards entrepreneurship.

Actually, we have followed this idea long before the recent proposals of the Ministry of Education. However, paths two and three are more challenging for us. We do have some practices such as ICT Showroom that support entrepreneurship stemming from innovations. The third path is the most challenging for us because we do not exercise basic research rather applied research and development. We work with existing research results and apply them to new environments and challenges.

Our entrepreneurship education is quite well structured, and it offers a different alternative to the students. Still, we think that we need to create more flexible ways to connect studying and entrepreneurship. We have called up a team focusing on the development of our entrepreneurship education.

This team has representatives from all our degree programs and we try to maintain the old and create new entrepreneurship education packages that combine students from different disciplines. Again, Practice Enterprise – the course is a good example.

Most of our students join this course during their first-year studies as suggested by the Ministry of Education, but engineering students join to this project in the third year. The student co-operatives answer to the inter-disciplinary requirement as well.

There are students from different degree programs, different disciplines working together for a common goal just like in working life. One of the areas that we have much to do is enhancing growth-oriented entrepreneurship.

In addition, we could take advantage of the large network of entrepreneurs and evolve them more in the planning and design of our entrepreneurship education. The goal of all our efforts in entrepreneurship education is not that everyone starts their enterprise.

We do hope that the number of startups of our students will increase. Hopefully, our efforts have given our student responsibility, understanding, and some inspiration.

REFERENCES

1. Cooney, T., (2012). *Entrepreneurship Skills for Growth-Orientated Businesses*. [eBook] Copenhagen, p. 23. Available at: http://www.oecd.org/cfe/leed/cooney_entrepreneurship_skills_HGF.pdf [Accessed 16 Aug. 2018].
2. Dunn, C., (2016). *12 Effective Ways to Build Entrepreneurial Skills That Matter*. [online] Entrepreneur. Available at: https://www.entrepreneur.com/article/279465 [Accessed 16 Aug. 2018].
3. *Entrepreneurs: Key Characteristics and Skills*. [eBook]. p. 6. Available at: http://www.cefe.illinois.edu/tools/Making%20A%20Job/MAJ_Student%20Guide%20Sample.pdf [Accessed 16 Aug. 2018].
4. *Focus on Entrepreneurial Skills*, (2014). [eBook]. p. 4. Available at: http://skillspanorama.cedefop.europa.eu/sites/default/files/EUSP_AH_Entrepreneurial_0.pdf [Accessed 16 Aug. 2018].
5. *Four Mini Case Studies in Entrepreneurship*, (2006). [eBook]. Higher Education Academy, p. 18. Available at: https://www.heacademy.ac.uk/system/files/4-mini-case-studies.pdf [Accessed 16 Aug. 2018].
6. Managementstudyguide.com. (2018). *Youth Entrepreneurship Building Skills*. [online] Available at: https://www.managementstudyguide.com/youth-entrepreneurship-building-skills.htm [Accessed 16 Aug. 2018].
7. Mindtools.com. (n.d.). *Entrepreneurial Skills the Skills You Need to Build a Great Business*. [online] Available at: https://www.mindtools.com/pages/article/newCDV_76.htm [Accessed 16 Aug. 2018].
8. Mukhuty, S., & Williams, T., (2015). *Entrepreneurship Skills: Literature and Policy Review.* Department for Business Innovation and Skills Research Report no. 236. [eBook] Research gate, p. 50. Available at: https://www.researchgate.net/profile/Sumona_Mukhuty/publication/281861542_Entrepreneurship_skills_literature_and_policy_review_Department_for_Business_Innovation_and_Skills_Research_Report_no_236_September/links/55fc0a4808aeafc8ac421176/Entrepreneurship-skills-literature-and-policy-review-Department-for-Business-Innovation-and-Skills-Research-Report-no–236-September.pdf?origin=publication_detail [Accessed 16 Aug. 2018].
9. Sousa, M., (2014). *Entrepreneurial Skills Development*. [eBook]. Research gate. p. 6. Available at: https://www.researchgate.net/profile/Maria_Sousa19/publication/268816221_Entrepreneurial_Skills_

Development/links/547880250cf293e2da2ad7a1/Entrepreneurial-Skills-Development.pdf?origin=publication_detail [Accessed 16 Aug. 2018].

10. Zimmerman, E., (2015). *The Skills That Make Entrepreneurs Extraordinary*. [online]. Stanford graduate school of business. Available at: https://www.gsb.stanford.edu/insights/skills-make-entrepreneurs-extraordinary [Accessed 16 Aug. 2018].

CHAPTER 6

E-AGRICULTURE

CONTENTS

E-agriculture, or ICTs in agriculture, is about developing, designing, and applying innovative ways to use Information and communications technologies in the rural domain, with a focus on agriculture. E-agriculture offers a wide range of solutions to some agricultural challenges. This chapter examined efforts taken by major developing countries to sketch the wide canvas of information and communications technology for agricultural developments.

This chapter will discuss the use of traditional information and communications technologies, such as off-line radio, TV, video, and printed newspapers. It will put more focus on the advent of information that is more recent and communications technologies devices and applications that are bringing new opportunities and challenges.

6.1. INTRODUCTION

E-Agriculture is an emerging field in connection with agricultural entrepreneurship, agricultural informatics and agricultural development referring to agricultural services, technology dissemination, and information obtain or improvised with the help of the Internet and related technologies. More precisely, e-agriculture comprises design, conceptualization, evaluation, development, and application of new and innovative ways to use existing or emerging information and communication technologies (ICTs).

The objective of E-Agriculture goes beyond technology; it helps to encourage the incorporation of technology with multimedia, culture, and knowledge, with the purpose of enhancing learning processes and communication amongst numerous actors in agriculture regionally, locally, and globally. Key components to e-Agriculture are:

- Facilitation;
- Technical support;
- Support of standards and norms;
- Capacity building;
- Education;
- Extension.

There are several types of activity related to e-agriculture applications that are widely recognized around the world today. The delivery of agricultural information and knowledge services (i.e. market prices, extension services, etc.) using the Internet and related technologies falls under the definition of e-Agriculture. More advanced applications of e agriculture in farming exist in the use of sophisticated information and communications technologies such as satellite systems, Global Positioning Systems (GPS), advanced computers and electronic systems to improve the quantity and quality of production (Figure 6.1).[1]

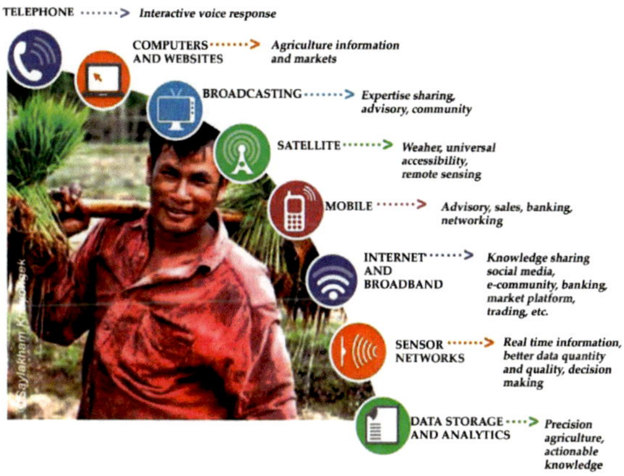

Figure 6.1. E-agriculture.

Source: https://i0.wp.com/www.ictworks.org/wp-content/uploads/2016/05/e-agriculture.jpg?fit=640%2C456&ssl=1

6.2. CURRENT SCENARIO OF AGRICULTURE SECTOR

The occupational structure of many developing countries is dominated by the "agricultural sector" and the "manufacturing sector" and the "service sector" is lagging far behind in this context. This shows that there is predominantly an agricultural economy and hence it needs the strongest development and protection of its "agricultural resources."

The world is facing some agricultural challenges that need to be resolved as soon as possible. The major challenges to the agriculture sector are:

- Insufficient agricultural infrastructure and support facilities.
- Insufficient institutional capacity to deliver farmers specific services.

1 World Summit on The Information Society, Geneva 2003 – Tunis 2005, Plan of Action, E-Agriculture A Definition and Profile of Its Application

- Lack of awareness about suitable agricultural methods among the farmers.
- Agricultural content development and its up gradations.
- Ownership issues of the public and government-generated data.
- Inadequate use of Public-Private Partnerships.
- Lack of "Common Platforms" for the farmers.
- The absence of an "Agricultural Think-Tank."
- Inadequate use of Information and Communications Technology for agricultural purposes, etc.

6.3. E-AGRICULTURE IN BRIEF

E-Agriculture Community comprises of individual stakeholders such as:

- Information and communication specialists;
- Researchers;
- Farmers;
- Students;
- Policymakers;
- Business people;
- Development practitioners.

More precisely, e-Agriculture encompasses the design, development, conceptualization, evaluation, and application of advanced methods to use information and communication technologies in the rural domain, while focusing on agriculture.

E-Agriculture comprises of internet platform of this global initiative intended to encourage sustainable agricultural development and food security by making appropriate the use of communication, information, and related technologies in the agriculture sector.

From farmers to researchers, e-Agriculture will connect all concerned persons together. The advantage of e-agriculture for farmers is that they can access the desired information at any time and anywhere in the world and farmers can get the help from experts seeing their problem directly without moving anywhere.

E-agriculture is an increasing field for improving present agriculture and food security activities through advanced processes for knowledge access by using information and communication technologies. E-agriculture is included in the World Summit on the Information Society (WSIS) Plan of Action in a region of the function of information and communication technologies.

The goal of the platform: In the field of agriculture and rural development practices e-agriculture is relatively a recent term. E-agriculture is a growing field focusing on the improvement of agricultural and rural growth with the help of advance information and communication processes. E-agriculture allows community members to exchange good practices, ideas, knowledge, and resources related to e-Agriculture, and to make sure that the knowledge created is successfully shared and used globally.

Architecture: An expert agricultural advice is necessary to the farmers both in a timely and personalized situation for improving agricultural productivity. Agricultural experts make the advice by using the modern agriculture in this system, which is extremely knowledge concentrated and requires timely, consistent, and precise information on natural resource endowments and their usage patterns at present and future technology available for their application and other information about markets, weather, insurance, subsidy, etc. The Architecture of the modern system is as follows (Figure 6.2):

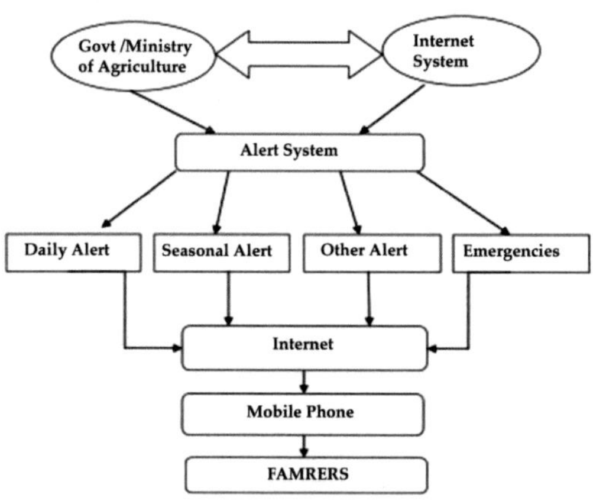

Figure 6.2. The architecture of the modern system.

Source: http://ijarcsse.com/Before_August_2017/docs/papers/Volume_5/1_January2015/V5I1–0292.pdf

6.4. GLOBAL TRENDS IN E-AGRICULTURE

6.4.1. Technology-Based Solutions

In intensive agricultural systems, applications of e-agriculture in developed nations are gearing towards using sophisticated technologies to improve the quality and quantity of production, to make the most of profits. In precision agriculture, farmers are connected to computer and satellite technologies to

- Improve yields;
- Cut their costs;
- Protect the environment.

E-commerce (or e-marketing) helps the entrepreneur to sell their agricultural products on electronic networks like the Internet and extranets. On the other hand, in many developing nations farmers can get information through grass root level initiatives of using information and communications technologies and from distance education modalities to improve the knowledge base among service providers.

6.4.2. Precision Agriculture

"In precision agriculture or site-specific farming, farmers are using information and communications technologies and other technologies to obtain more precise information about agricultural resources which allow them to identify, analyze, and manage the spatial and temporal variability of soil and plants for optimum profitability, sustainability, and protection of the environment."[2]

Precision agriculture is described as "A system to manage farm resources better. Precision farming is an information technology-based management system now possible because of several technologies currently available to agriculture. These include global positioning systems, geographic information systems, yield monitoring devices, soil, plant, and pest sensors, remote sensing, and variable rate technologies for application of inputs."[3]

Precision agriculture is an advanced e-agriculture application. It makes use of five major components of technology:

---

2 Jefferson County Cooperative Extension Services. 2000. Precision Agriculture-Site Specific Farming

3 E-Agriculture and Rural Development: Global Innovation and Future Prospects-Blessing Maumbe And Charalampos Z. Pratikakis, Dec2012

- Geographical Information Systems (GIS) for analysis and management of spatial data and mapping.
- Remote Sensing (RS) to identify.
- Global Positioning Systems (GPS) to locate and define spatial features or activities which contribute to the quality of site-specific practices.
- Variable Rate Technology (VRT) allowing targeted, site-specific input applications.
- Yield observation for recording crop productivity as a historical database for crop management.[4]

E-Commerce brings the biggest advancement in agriculture and helps in improving productions and high yields result and create a profitable market outside local communities. Electronic markets are offering a global platform to farmers to sell and market their products to buyers at the global level. Electronic commerce (e-commerce) is generally defined as the exchange of goods and services over the internet, and e-commerce is already having an important impact on agriculture.

Farms had already bought or sold agricultural products on the Internet[5] and Goldman Sachs had estimated that 12% of all agricultural sales in the U.S. would be conducted over the Internet in 2004, compared to only 4% in 1999.[6] Further, a study conducted by Rockwood Research on Internet use by commercial farmers in the United States found that farmers were primarily using the Internet to access information on:

- Machinery;
- Commodity prices;
- Farm chemicals;
- Weather.

The study also showed that farmers were migrating quickly toward Web-based transactions such as to buy farm equipment, seed, and crop chemicals on the Internet.[7]

---

4 Rains, G. C. and Thomas, D. L. (2000(. Precision Farming: An Introduction.

5 U.S. Department of Agriculture, Economic Research Service, Agricultural Resource Management Study, 1999.

6 Rolf A.E. Mueller. (2000). Emergent E-Commerce in Agriculture. Aic Issues Brief, No. 14, December.

7 Farmers Go Online in the US – Fast. Online Publishing News, No. 16, 5 July 1999.

6.5. E-AGRICULTURE AND EXPERIENCES OF ENTREPRENEURS IN AGRICULTURE

A huge population in the rural areas globally is dependent on the agriculture and agricultural activities for their livelihood. Approximately half of them are in the poor category, and most of the money they earn is spent on completing their basic needs, particularly the food needs of the family.

With the increasing population over the last few decades, their per capita share of land and water resources has diminished substantially. As a result, the people in rural areas in form of unemployment face the major troubles.

As agriculture is one of the major sources of income and livelihood, the profitability along with the productivity in the agriculture sector has been considerably low in many of the nations. While the increasing health problems and illiteracy can be seen as the reason for falling productivity, poor dissemination of technology, lack of awareness, poor communication and information services, inadequate investment in agricultural inputs, degradation in profitability can be attributed to inefficient and inadequate infrastructure that is required for backward and forward integration, poor post-harvest, and poor connectivity with market, poor post-harvest and processing facilities, that ultimately results in exploitation by large number of middlemen.

In older times, the dependency of farmers was majorly on the Agricultural Extension Agencies of the State Government for information, input supply, and marketing services. Over the course of time, these bodies are not being efficient and able to keep up the pace with the growing responsibilities and specific need of the farmers.

To overcome this issue, cooperatives by farmers have been promoted to supply various agricultural inputs, manage, and organize the marketing needs of the farmers and provide a right market for farm produce. While, this was a good approach but most of the cooperatives were not able to carry out the work in an efficient and economic manner due to the lack of commitment from the elected leaders and unfair practices in trading by competing traders. The farmers and producers were not realizing the inability of the extension agencies, are still dependent on the external and government agencies to help them in managing their agri-business.

In most of the regions on the global platform, agriculture is considered as a family tradition and many of the farmers continue to practice what their ancestors and neighbors practiced. This profession is now asking for the reforms. Agriculture as a profession must be considered as an enterprise, which should be supported with sound management back up.

As compared to any other profession or enterprise, there is a need for of proper planning regarding demand forecast, inventory of resources, choice of technology, the skill level of the available human resources, need for external inputs their training needs, infrastructure, and services needed for carrying out various operations and marketing. The primary step is the change in the mindset of the Agricultural extension agencies and farmers for promoting successful entrepreneurship in agriculture (Figure 6.3).

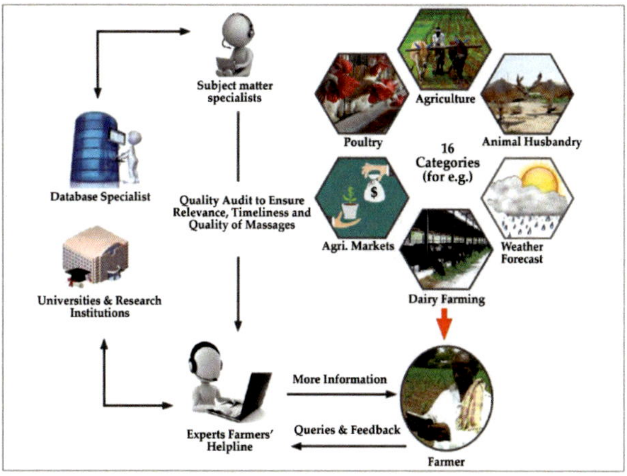

Figure 6.3. Process of digital agriculture.

Source: http://www.icrisat.org/digital-agriculture/

6.6. ENTREPRENEURSHIP IN AGRICULTURE

In the lack of local entrepreneurship, the opportunities in the field of agriculture are overtaken by people from outside, especially the businessmen and traders from urban locations, leading to deprivation and exploitation of employment of the farmers.

Considering the ever-increasing unemployment in rural areas and slow growth rates of the agriculture sector, it is important to explore all the options for the development and promotion of entrepreneurs in agriculture, which in place can address the present issues and troubles related to agricultural profitability and production.

6.7. TYPES OF ENTERPRISES

There are different types of enterprises that exist in agri-business and play a great role in the promotion of entrepreneurship

- **Farm Level Producers**: at an individual family level, each family is to be considered as an enterprise, to optimize the production by using the technology, demand in the market and resources at the best.
- **Providers**: to optimize agriculture by every family enterprise, there are several types of services that are necessary and required at the grassroots level. These include
    - The input distribution and procurement;
    - Hiring of implements and equipment like seed drills, tractors, harvesters, sprayers, dryers, threshers;
    - Technical services such as weed control, installation of irrigation facilities, transportation, storage, plant protection, harvesting, threshing.

Similar openings exist in the livestock and husbandry sector for providing vaccination, breeding, treatment services, and disease diagnostic apart from distribution of mineral mixture, cattle feed, forage seeds, etc.

- **Input Producers:** there are many prosperous enterprises, for which important input skills are needed. Some of the inputs that can be given by the entrepreneurs at ground levels or village level or local level are root media for raising plants in pots, biopesticides, plants of different species of fruits, vegetables, mineral mixture, irrigation accessories, production of cattle feed concentrate, ornamentals, soil amendments, vermicomposting, agricultural tools, bio fertilizers, complete feed. There are good chances to support fishery, sericulture, and poultry as well, via promotion and facilities in critical service in rural areas.
- **Marketing and Processing of Farm Produce:** Well-organized processing and management of post-production processes need investment as well as the higher scale of technology. Such processes and enterprises can be managed by People's organizations, either in the form of service societies, cooperatives or joint stock companies. The most successful examples are the dairy cooperatives, sugar cooperatives, and fruit producer cooperatives in many countries. However, the rise of this kind of ventures completely depends on the competence and integrity

of the leaders who are leading the movement. Good professional support is required for this kind of venture and for managing the activities in form of competitive business and to contest in a competitive spirit with other players in the market, particularly the intermediaries and retail traders.

6.8. PROBLEMS FACED DURING DEVELOPMENT OF ENTREPRENEURSHIP

- Agriculture and entrepreneurship go hand in hand, as it is not only a chance but also a requirement for improving the profitability and production. However, the rate of success is very low. The following reasons have been brought into the spotlight:
- For many of the farmers, agriculture is just a way to survive that fulfills their minimum basic needs. In the lack of proper information and knowledge, technology, resources, and connectivity with the potential market, it is hard for the illiterate farmers and small stakeholders to convert their agriculture into a well-organized enterprise.
- Before broadcasting and promoting different services by self-employed personnel, it is required to generate awareness and make them informed about latest upgradations taking place among the farmers, who are the ultimate users, regarding the advantages of these services.
- To publicize the services, the recent actions like supplying service free of cost by the Government and local agencies should be discontinued. In fact, most of the farmers, especially the ones who have strong political connections and have links with local leaders are under the impression that the government is responsible for providing technical support and extension advisory services to the farmers.

However, over the time, the trustworthiness has battered and the outcome was that the services provided by these agencies are not extended to small farmers, especially those who belong to remote areas. Nonetheless, the notion of free service is making the farmers unwilling to avail of paid services, offered by the local self-employed technicians.

- The technicians who are self-employed require regular backup services in the form of business information and technical help, contact with marketing agencies, equipment, and research stations, suppliers of critical inputs and who are involved in the development of modern technologies.

- Many legal obstacles and restrictions create hurdles in the progress of agricultural business, and are promoted by the People's Cooperatives and organizations. Private traders are involved in the types business tends to ignore rules and regulations and create disturbance in the fair-trade environment.
- Organizations by the public often keep a back foot in taking the risk of making drastic changes like adopting modern and expensive technologies and making huge investments, which in long-term affects their profitability. In addition, with outdated machinery and technology, along with low profitability, farmers often drop the interest in their own initiatives as well as in their leaders.

6.9. STRATEGIES IMPLEMENTED FOR PROMOTION OF SUCCESSFUL ENTERPRISES

- Considering the current problems and hurdles faced by the entrepreneurs involved in agri-business, it is crucial to creating a friendly environment in the field. Some of the required conditions that play an important role in the success of agri-business are mentioned below: There shall be common and viable option for the officials working in the governmental organizations and farmers about the benefits and need of promoting youth and private who are self-employed entrepreneurs to make enable the farmers to enhance agricultural profitability and production.
- The government should completely halt the practices of providing free services in the sectors where the private entrepreneurs are more active and are working in public interest.
- The technical ability and skills of the entrepreneurs should be calculated to ensure high standards. There should be a watchdog to do proper monitoring and surveillance to keep a check on the quality of services and the charges collected from the small farmers to overcome the exploitation they face.
- The government bodies should motivate and encourage the entrepreneurs with the introduction of various concessions and incentives.

6.10. ROLE OF INFORMATION AND COMMUNICATIONS TECHNOLOGY IN E –AGRICULTURE

Information and communications technology comprise of communication device or application, electronic devices like radio, television, smartphone, computer enable network hardware and software, satellite systems and numerous services and applications related to mention devices, like video conferencing, distance learning necessary for the delivery of information in the form of text, audio, video, data, image etc. from one point to another.

Information and communications technology involve all technical means used to manage information and aid communication. Significance and extraordinary information and communications technology productivity gains are not only for individuals and businesses, but also for a country.

E-Agriculture is a new concept about agricultural informatics, which has made rapid development in information and communication technologies and of the internet. Agricultural informatics is an evolving field that integrates the advances in agricultural development, informatics, and entrepreneurship to offer:

- Advance technology dissemination;
- Better agricultural services;
- Delivery of information through the advances in information and communications technology and the internet.

The broadcasting of information to farmers has become progressively integrated into information and communications technologies. Information on education, agricultural, and health issues is provided by rural telecentres. In addition, rural telecenters provide basic literacy and skills to use computers to rural citizens and they broadcast agricultural information through radio and television programmers.

Transfer of technology in the agricultural sector is facilitated by many of the organizations like private, government, cooperatives, and the public. Information and communication technologies are important while enabling communication and access to information for rural and agricultural development. Due to information and communication technologies wide application and appeal, it marks a tremendous impact on the rural economy.

Advantages of information and communications technology in E-Agriculture

- Information and communication technology can help in initiating new agricultural and rural business like:

- E-commerce;
- Rural tourism;
- Real estate business for satellite offices;
- Virtual corporation of small-scale farms.

Information and communication technology can support policy-making and evaluation with the help of tools like geographic information systems (GIS) on

- Disaster management;
- Optimal farm production;
- Agro-environmental resource management, etc.

Information and communication technology can enhance farming technologies and farm management, understanding competitive and sustainable farming with safe products by efficient:

- Farm management;
- Risk management;
- Effective information or knowledge transfer etc.

Information and communication technology provides direction to farmers on:

- What to plant?
- When to plant?
- How to manage pests?

Along with considering off-farm factors such as:

- Environmental impacts;
- Industry standards;
- Market access.

IT-based decision support system (DSS) can surely help their decisions. It can provide systems and tools to secure food traceability and reliability that has been an emerging issue concerning farm products since serious contamination such as chicken flu was detected (Table 6.1).

1. Information and communication technology can enable rural activities and offer more easy, comfortable, and safe rural life with equivalent services to those in the urban areas, such as:
    - Telemedicine;

- Provision of distance education;
- Remote public services;
- Remote entertainment etc.

2. Empowerment of Stakeholders (Government Officials, Research, Education & Extension Scientists, farmers, and other service providers such as Community Information centers).
3. Efficient management (Development, Conservation, allocation, and utilization) of resources.
4. Development of Knowledge Management, Decision Support, and Advisory Systems to strengthen extension services and used for farmers redressal system.
5. Improved productivity and profitability of farmers through better advisory systems.

Table 6.1. E-Agriculture Vision for Farmers

Service	Current situation	Situation by 2020 with the delivery of the e-agriculture vision
Accessibility to information and services	Not accessible by all, particularly in remote areas	Accessible by all
Timeliness of services	Timely services not guaranteed	Timely services available to the farming community
Production system services	Incomplete and generalized	Holistic and site-specific resulting in increased efficiency of inputs and saving of cost
Pest and disease outbreaks	Incomplete and untimely information and inefficient way of managing a crisis resulting in huge economic loss	Better management of pest and disease outbreaks with complete and timely information resulting in crop failure prevention
Weather information	Incomplete and generalized information resulting in poor benefits	Holistic and site-specific information resulting in better benefits
Market access	Largely limited to information on commodity prices	Holistic information on markets, prices, transport, storage. Trading with less or no involvement of intermediaries or agents

Payments and money transfers	Largely done manually with cash – slow, tedious, inefficient, and involving considerable cost and risk	Fast, transparent, and reliable money transactions at affordable charges with reduced risk
Interaction with experts	Limited	Significantly improved level of direct interaction with experts

Source: http://www.fao.org/3/a-i5564e.pdf

6.11. E-AGRICULTURE OPPORTUNITIES

E-Agriculture projects provide opportunities and benefits to a wide range of stakeholders, from small farmers to businesses and governments.

Entrepreneurs

- New job opportunities with growth potential.
- Opportunities to deliver much-needed services to farmers (soil testing, financing, logistics, farm machinery, etc.).

Farmers

- Lower costs.
- Increased productivity and income.
- Better risk mitigation (e.g., pests, disease, and weather).
- Greater access to expert information, services, and best practices.

Governments

- Economic growth and job creation in rural areas.
- Greater food security.
- Improved relationships with rural communities.

Market aggregators

- Ability to directly and cost-effectively source produce in bulk.
- Ability to optimize market efficiency by connecting buyers and sellers.

Microfinance institutions

- Access to detailed information on farmers to better assess credit-worthiness and manage risk.
- Ability to serve new customers who previously had no formal access to affordable finance.

- A better understanding of unmet financial needs, enabling the creation of innovative financial products such as crop loans and insurance.

Suppliers

- Direct, inexpensive, and transparent access to larger groups of farmers.
- Better forecasting for greater efficiency and fewer losses.
- Stronger relationships with farmers.

Technology companies

- New business opportunities using a proven model that increases readiness for technology adoption.
- Ability to reach new customer base early, for longer-term benefits.
- Opportunity to achieve business and social benefits by increasing access to technology.

Universities and extension offices

- Direct access to otherwise unreachable farmers.
- Ability to test and validate academic research and innovation in real-world applications.
- Opportunity to solve practical problems and share best practices with aggregated groups of farmers.
- Enhanced student learning through hands-on projects.

Reaffirming the Relevance of information and communications technologies for Agriculture and Youth

This section clarifies the relevance of information and communications technologies in African agriculture and the status of its use in the sector. In addition, it discusses the rationale underlying the use of these technologies by youth in agriculture.

6.12. THE DIVERSITY OF INFORMATION AND COMMUNICATIONS TECHNOLOGIES FOR AGRICULTURE

Information and communications technologies have conventionally been "based on indigenous forms of storytelling, song, and theatre, print media and radio." With the arrival of modern information and communications technologies, specifically, the Internet and the mobile technology has transformed the way data/information is shared and the speed of communication all over the continent.[8]

Information and communications technologies have also decreased the cost of accessing new knowledge and information and is creating many new opportunities in different sectors of African economies, including agriculture.[9]

Traditional media, such as television, radio, print, and video are still appropriate communication channels for a country like African agriculture. Apart from email and websites, the use of which is becoming commonplace, even in the agricultural sector, modern information and communications technology devices and applications used in agriculture now include:

- Computers;
- Tablets;
- Mobile phones;
- Television;
- Satellites;
- Office software;
- Short Messaging Services (SMS);
- Social media;
- Geographical Information Systems (GIS);
- Drones.

The use of which is being explored in some African countries, such as Ghana (Newsghana, 2015).[10]

Not all forms of information and communications technologies are yet fully applicable to agriculture in the African context. Some are more relevant than others based on factors like cost, accessibility, applicability, user profile, and so on.

Below is a brief discussion of a few prevalent information and communications technology devices and applications:

- **Mobile phones and SMS:** The mobile phones are now widely used in Africa. The SMS is a messaging service of the web or mobile phones that enables users to exchange short text messages

8 Conway, 2012

9 Juma, 2011

10 http://www.detroitnews.com/story/business/2015/07/06/farm-use-drones-take/29792757/

with their devices. SMS is hugely popular, in Africa and globally. Much innovative information and communications technologies for agriculture in Africa are SMS-based.

- **Online TV and videos:** Videos have been traditionally used to disseminate agro-advisory information, if required in local languages, through a variety of channels. Online videos can be watched on a computer, a tablet, or a mobile phone via websites like YouTube and Access Agriculture, provided there is good connectivity. Dedicated TV programs for young farmers, such as the 'Shamba Shape-Up' TV show in Kenya, are available in some nations.
- **Online radio:** The radio is perhaps one of the most effective ways to pass information (notably in local languages) to farmers and rural households in most developing countries, whether in Africa or elsewhere – provided the content of the message be packaged in an appealing way.[11] "Agricultural audio files and programs can be downloaded from the Internet on many websites. Similarly, some community/rural radio stations that broadcast agricultural programs (among others) are providing online streaming."[12]
- Office software;

  Office software refers to applications used for analyzing data and text processing, such as:

  - Microsoft's Word and Excel programs;
  - Apple's Pages and Numbers;
  - OpenOffice Writer and Calc.

For young farmers, text editing software and spreadsheets are particularly useful due to their wide availability. This software is embedded in new computers, is free of charge, and is essential for efficiently processing production data and finances. Still, to use these basic tools many young farmers lack the literacy in rural areas.

- **Social media:** "These can be defined as electronic information and communication platforms that enable users to easily create and disseminate content on digital networks and engage in interactive communications."[13] Facebook and Twitter-like social media platforms are very easy to operate and have become

11 Chapman, et al., 2003; Farm Radio International, 2008; Nakabugu, 2001
12 Ndiaye and Lohento,2008
13 Kaplan and Haenlein, 2009

effective means of information-sharing tools for tech-savvy farmers, particularly youth, as they are very active on these social media platforms.

- **Blogs:** Blogs are now widely used as key external institutional communication platforms, and are easy to write and published on freely available platforms as agricultural institutions cannot pay for the time or cost of creating a proper website. Such social media tools are increasingly used as business promotion platforms (for example, in the marketing of agricultural products).
- **GPS and GIS:** Geographic information systems (GIS) enable the capture, analysis, management, and display of geographical and spatial information.[14] They can include the use of GPS (Global Positioning System) receivers, which connect to satellites to facilitate access to locations. "The GPS has been notably beneficial to agriculture in monitoring climate variability, and for providing current weather forecasts for early warning systems that guide decision-making by farmers and relevant government agencies."[15] GPS technology is now commonly embedded in digital cameras and mobile smartphones (Table 6.2).

Table 6.2. Examples of Common E-Agriculture Infrastructure Components

Components	Description	Examples
Communication infrastructure	Communication infrastructure coverage that enables people to access e-agriculture services using voice, SMS, MMS, social media, radio stations, sensor networks, and television.	• Mobile, fixed, and satellite networks; • Cloud networks and locally-hosted data storage; • Radio stations and towers; • Television towers; • Sensor networks; • Interactive voice response systems (IVRS); • Data analytics services.
High-speed data connectivity	The high-speed network connectivity infrastructure required to access advanced e-agriculture services and applications.	• Submarine communications' cables; • Broadband connections; • Internet.

14 Maliene, et al., 2011

15 Pearson, 2012

Electrical infrastructure	By its very definition, e-agriculture requires access to electricity at some point in order to function.	• Power plants; • Power lines; and • Off the grid power (e.g., solar panels).
Computing infrastructure (hardware and software)	The physical computing infrastructure (e.g., PCs, laptops, mobile phones, smartphones, tablets, server infrastructure, sensors, and associated networks, switches, etc.) and the software applications which enable the collection, recording, and exchange of electronic information across the agriculture sector, and support agricultural service delivery.	• National, state, regional, and local service and computing infrastructure; • Agricultural service provider computing infrastructure; • Financial service provider computing infrastructure (switches, ATMs, POS devices, etc.); • Smartphone and desktop applications; and • Applications platforms e.g., Content Management Systems.
Agricultural content, datasets, and related applications	Datasets that support agricultural service management and administration, which typically provide access to longitudinal and aggregated information for analysis, reporting, research, and decision-making. Databases and content repositories that enable farmers and agricultural service providers to access trusted and verified agricultural information and knowledge.	• Weather, pest, disease, price, production, and other national, regional, and local databases; • Open access to data and big data analysis; • GIS and satellite data; • Disaster-related data; and • Agricultural knowledge resources.

Source: http://www.fao.org/3/a-i5564e.pdf

6.13. RECENT DEVELOPMENTS AND EMERGING TRENDS

The sector of information and communications technologies for agriculture and rural development established the new trend to important areas for highlighting the following opinions:

- **Mobile telephony:** In the past ten years, there is a drastic change in the mobile telephony penetration. The increment is premeditated and this led to an increase in mobile applications, which specifically designed for agriculture and rural development. The number of mobile platforms developed and accessible are used in the market to conduct the digital device with slight landholder, farmers have also increased enormously, bringing a more assorted range of information sources for farmers (e.g., Internet, radio, TV, newspaper, and extension agents) than a few years ago.

Mobile telephony provides available and relevant information that distributes and holds great potential and is being well thought-out or it might be used as an important channel for agricultural advisory services, financial services, and other essential information in rural communities, especially in enhancing access to markets, information on market supply and demand.

- **From mobile phones to smartphones:** A few years ago, Short Message Service (SMS) dominated, now there are:
  - IVR (interactive voice response);
  - Smartphone apps;
  - App integration with social media.

Young people have driven the development of mobile applications (apps) for agricultural development partially. Apps alone may not cater to the needs of farmers, but they are a huge step towards integrating agriculture and ICTs.

- **Mobile financial services:** There are four types of financial services that farmers require to reach their economic goals have been recognized and are becoming more widespread:
  - Credit;
  - Savings;
  - Money transfers and payment facilities;

Insurance.

Information and communications technologies have created the potential to offer various financial products to a large number of rural clients than traditional financial service providers are not able to reach so far. Information and communications technologies can help to improve economic efficiency and improved service delivery, as they lower business and transaction costs.

- **Use of information and communications technologies in agriculture:** Information and communications technology innovation play an important part in improving the value chain and agricultural production. Food traceability systems with the implementation of information and communications technologies have become very important risk management tools that allow food business authorities and operators to cover food safety problems and encourage consumer confidence.

Information and communications technology play a major role in enabling marketing and entree to markets, particularly to access data on market demand and market prices. Information and communications technologies enhanced marketing and certification strengthens. The capacity of small-scale producers to enhance their revenue by establishing their situation on local as well as international markets. Agro-meteorological technologies and GIS are been implemented into programmes for numerous purposes including:

- Crop forecasting;
- Land-use planning;
- Early warning systems.

To observe threats from the rising number of natural disasters space technology is also essential. Furthermore, the use of mobile phones has become more common for exchanging information on pest tracking and disease surveillance. There is also an increasing prevalence of information and communications technology solutions for the later stages of the agricultural value chain. For example:

  - Post-harvest;
  - Transport;
  - Storage.

- **E-Agriculture strategies in information and communications technology policies:** There are no specific information and communications technologies for agriculture strategies in many countries but efforts are made in e-agriculture strategy initiatives in a few nations like Ghana, Côte d'Ivoire, Mali, and Rwanda. Furthermore, in most information and communications technology policies established with the support of organizations like ITU, UNECA, UNDP, and IICD, there are provisions on sectoral strategies for agriculture. However, approaches vary in different nations and regions.
- **Online learning:** A facility of new information and learning to smallholder farmers can be provided through open online

courses, and their combination with educational content and scientific content that is more broadly available (even on social media). With online learning, it possible to observe the capacity of communities and institutions to use this information effectively.

- **Growing use of big/open/real-time data collection and analysis:** Use of big/open/real-time data is an innovative application used in decision support and farm management, but also increasing issues of intellectual asset management, mainly when data are captured at the farm level. Equally applicable to local adaptation of content is the approach based on content co-generation (Table 6.3), which in turn increases issues of governance of:
    - Information;
    - Data;
    - Skills and technology;
    - Development of open standards and technologies.

Table 6.3. Examples of Governance Functions, Responsibilities, and Composition

Group	Responsibilities	Composition
Agriculture sector leadership in collaboration with the ICT sector	• Gives overall direction, oversight, and mandate; • Secures spending authority and resources; • Acts as the vocal and visible champion; • Assists with resolution of major issues, problems, conflicts, and other challenges; and • Approves, endorses, and owns the national e-agriculture vision.	Senior-level agricultural and ICT sector decision-makers, such as the Minister or a senior bureaucrat in the Ministry of Agriculture and Ministry of ICT and/or regulatory authority.

Steering Committee	• Acts individually and collectively as a vocal and visible champion through its representative organizations; • Provides direction and guidance to other groups; • Plans and manages the vision development process; • Makes decisions at key stages of the process; • Organizes information gathering, analysis, and drafting of the national e-agriculture vision; Assists in addressing risks, resolving issues and conflicts; • Oversees overall progress, and approves changes to scope or approach; and • Provides support in policy and advocacy.	Those individuals who should be involved in making decisions in relation to the development of the national e-agriculture vision, the acceptance of the vision, and the progression of its recommendations, including representatives from ministries and sector regulators (agriculture, ICT, finance, commerce, governance, meteorology, etc.). Representatives of the industry could be considered.
Subject-matter experts and advisory groups	• Provide guidance on the development of outcomes and recommendations, and support the development of the national e-agriculture vision; • Assist in the identification of existing or planned e-agriculture components, and their re-use or sharing; • Provide recommendations on specific reforms; • Provide insights into the implications of strategic directions and recommendations for the stakeholder groups; • Review and provide feedback on findings, conclusions, and draft deliverables; • Help in service design and define user experience according to the target group; and • Define information management standards (such as entity definition, metadata standards etc.).	Includes input from academics, thought leaders and representatives from the agriculture sector, ICT sector, other concerned sectors, and farmers' organizations. Not involved directly in decision making but able to exert a high degree of influence due to their acknowledged expertise in the field and/or their role as formal or informal advisors to key decision makers.

Broad stakeholder environment	• Responsible for soliciting broad stakeholder input to the development of the national e-agriculture vision; • Piloting of proof of concept; and • Provide feedback on deliverables that have been shared or published.	Involves input from individuals and organizations that are impacted by, or have a particular interest in e-agriculture and the outcomes of the process.

Source: http://www.fao.org/3/a-i5564e.pdf

6.14. FUTURE SCOPE

- For the young farmer e-agriculture is very beneficial as it offers essential information's about the plantations that they have grown.
- E-agriculture scheme using the data mining technique namely birch clustering has been used for clustering the large datasets of farmers details. The present work on E-agriculture conveys the information regarding agricultural details to farmers in SMS via SMS gateway.

The details such as daily alert, seasonal alert, and other additional details can be sent to farmers. The daily alert can be sent to all farmers in the database. The seasonal alert can be sent to farmers only for selected farmers based on the clustering result. Finally, the other or additional detail, which is announced by agriculture, can be sent to all farmers. The experimental result shows the better result when compared with the existing work

- This chapter has examined efforts taken by major developed countries to sketch the wide canvas of information and communications technology for agricultural developments. This is then thought in the lines for the potential benefit of Indian agricultural developments and rural developments in general.
- Majority of farmers in the state or country are aware that mobile phones can be used to conduct businesses and receive information. Mobile phone costs should be lowered to enable the majority of farmers for having access to the current information about agribusiness within the state or country.
- The government should also conduct sensitization to create awareness for the farmers on how best they can use information technologies to conduct agribusiness.

6.15. CONCLUSION

The present work on E-agriculture conveys the information regarding agricultural details to farmers in SMS via SMS gateway and hereby proposes to switch over E-agriculture. The details such as daily alert, seasonal alert, and other additional details can be sent to farmers.

The daily alert can be sent to all farmers in the database. The seasonal alert can be sent to farmers only for selected farmers based on the clustering result. Finally, the other or additional detail, which is announced by agriculture, can be sent to all farmers. The experimental result shows the better result when compared with the existing work. This paper also talks about Pros and cons of E-agriculture.

6.16. CASE STUDY: HARTMANN FARM, SZÁKSZEND VILLAGE, KOMÁROM-ESZTERGOM, HUNGARY

The entrepreneurial capacity of the next generation allows investment in technologies such as precision farming, which can facilitate the sustainable intensification of agriculture.

In Brief:

- Estate of c. 500 Ha
- 385 Ha owned by the family; 120 Ha rented from other landowners
- Hartmann Farm is owned and managed by Imre Hartmann, and his son-in-law, Andràs Takács
- Arable farm: Maize, Wheat, Sunflower, & Rape Seed
- Since 2009, Andràs Takács has been investing in technology for precision farming

Family History: Imre Hartmann has been involved with agriculture for all his working life. During the communist era, he headed one of the sectors of a co-operative farm. As an independent farmer, he started rearing Hungarian Grey Steppe Cattle in the 1980s and pork later in the 1990s.

After the Restitution, Imre quickly saw the value of agricultural land, and began to buy what he could. By 1993, he had a farm of between 50–60 Ha. In 2001, his son-in-law, Andràs Takács joined him in working on and managing the Hartmann farm, now consisting of some 500 Ha. Andràs himself owns 100 Ha of this land, which he manages as a young farmer.

Structure, Management & Labor: Hartmann Farm is divided into three legal entities, the Hartmann Farm Kft Company; the Hartmann Family Farm; and, Takács Andràs Young Farmer enterprise. Andràs Takács and his father-in-law, Imre Hartmann, manage the company and family farm jointly. Imre's daughter and Andràs's wife, Anita Takácsné Hartmann is the office manager for Hartmann Farm Kft, and does the daily office work as well as the bookkeeping. Her brother, who works in the financial sector, provides further support in this respect.

In addition to the family members, two permanent external employees work on the Hartmann Farm.

Agriculture & Environment: At present, 4 crops are rotated on the Hartmann enterprise, Maize, Wheat, Sunflower, and Rape Seed. The location of the farm, in the Kisalföld region in the North West of the Country provides optimal conditions for Maize cultivation, which represents 80% of the area's agricultural production. In line with new EU recommendations, this rotation will be reduced to three, with Sunflower Seeds gradually omitted from cultivation.

Products are stored onsite and then sold on a contractual annual basis to a regional buyer, providing market security. Indeed, while the farm has had to economize in the current economic climate, it has withstood the crisis well. In 2013, wheat was supplied from the Hartmann farm to the large national pasta factory, Gyermelyi, and was also exported to Italy.

Imre and Andràs practice conventional agriculture and rely upon crop rotation as well as the application of pesticide to avoid infestations of Western Corn Rootworm. However, in December 2013, the region was declared 'Nitrates Sensitive,' thus the extensive use of fertilizer will likely be limited in the future. They have been practicing minimum tillage for 4 years and are really seeing the positive impact on yields. In addition, the Hartmanns have a share in a collectively owned, externally managed Natura 2000 site.

Young Farmer Entrepreneurship: In 2009, Andràs won a call for tender for New Young Farmers. This gave him the capital to apply for a loan enabling him to invest in a GPS. Since then, Andràs has been slowly building up the techniques of precision farming on his own 100 Ha plot of land. A robot pilot guides Andràs's tractor to sow and harvest in perfectly straight lines, while computer and GPS sprayers control the application of crop protection products and water. This investment in technology for increasing the efficiency of agricultural output by avoiding any excess of environmentally detrimental inputs is extremely important for the future. It is a way of optimizing inputs and outputs in the drive for the sustainable intensification of agriculture. Andràs exemplifies the dynamic entrepreneurship that comes from young farmers and the next generation. He is an active member of AGRYA, the Hungarian Association of Young Farmers.

Community Leaders: The entire family plays an important social role, has extensive knowledge of, and contacts throughout the county. Imre is currently sitting as Deputy Mayor for the fourth consecutive term. Anita organizes a local wine contest and charity ball, as well as PIONEER communication days and visits. In 2011, Andràs won an award for the County's best Agricultural Enterprise.

6.17. CASE STUDY: TINAJEROS, ALBACETE, LA MANCHA, SPAIN

The economy of scale on the Tinajeros Farm allows it to make a highly significant contribution to rural employment and rural development. It permits investment in technology to advance and improve the farming, but also an investment in people and place.

In Brief:

- 5650 Ha Farm in semi-arid La Mancha region.
- 500 Ha Irrigated Agricultural Land; 400 Ha Non-Irrigated Agricultural Land; 100 Ha Grassland; 4650 Ha grazing for sheep and dairy cattle.
- Limited Liability Company, with ownership, split 4 ways between a mother and three sons.
- The business employs 35 external, permanent workers.

History, Management and Employment: This very large family farm and business near Albacete, Spain was brought together to form a Limited Company with the marriage of the parents of the current part-owner and manager. The property on the paternal side of this arrangement has been within the same family for 200 years.

Today the estate is split four ways between a mother and her three sons. The eldest son manages the farm, while the other family members contribute to farm work on an ad. Hoc basis. Due to its size, the farm can provide specialized and unskilled permanent employment for 35 individuals. Four further workers support an agricultural engineer; the director of the dairy relies on a staff of 10; and, 10 workers support the sheep-herding director. Three office staff is employed, in addition to a general director, and three further temporary farm workers are on hand for busy periods.

Regional Development: Community Engagement: The Tinajeros farm thus makes a highly important contribution to rural employment in the region. This has forged a strong relationship between the family and the local community through the workers, but also with suppliers and transporters who stay and eat in the neighborhood when picking up products.

Investing in Technology: Tinajeros makes a strong annual return on agricultural products of c. 4.5 million Euros. This allows for an investment of between 5 and 10% each year in the latest technology and innovations to ensure optimum yields with the most efficient use of inputs. The farm uses GPS facilitated computers to control tractors and irrigators, thereby avoiding waste of water and electricity. They have recently invested in solar panel technology for lighting, and in new cooling systems for the milk from the dairy.

In addition, this commitment to technology is a commitment to quality. Tinajeros supplies its best quality lamb and milk products directly to small local butchers, and cheese and yogurt factories.

REFERENCES

1. Bailey, E., (2014). *Families – The Entrepreneurs of the Countryside*. [online] Europeanlandowners.org. Available at: https://www.europeanlandowners.org/files/pdf/2014/family_farming_brochure_250x150cmyk_3mmbleed.pdf [Accessed 18 Aug. 2018].
2. Fao.org. (2015). *E-Agriculture Strategy Guide*. [online] Available at: http://www.fao.org/3/a-i5564e.pdf [Accessed 18 Aug. 2018].
3. Ghogare, S., & Monga, M. P., (2015). *"E-Agriculture" Introduction and Figuration of its Application*. [online] Ijarcsse.com. Available at: http://ijarcsse.com/Before_August_2017/docs/papers/Volume_5/1_January2015/V5I1–0292.pdf [Accessed 18 Aug. 2018].
4. Itu.int. (2014). *E-Agriculture*. [online] Available at: https://www.itu.int/net/wsis/review/inc/docs/ralfreports/WSIS10_ALF_Reporting-C7_E-Agriculture.Summary.pdf [Accessed 18 Aug. 2018].
5. Lohento, K., & Ajilore, D. O., (2015). *ICT and Youth in Agriculture*. [online] Available at: https://www.researchgate.net/publication/308968253_ICT_and_Youth_in_Agriculture [Accessed 18 Aug. 2018].
6. Rahman, R., & Fong, J., (2016). *Innovate for Agriculture*. [online] Publications.cta.int. Available at: https://publications.cta.int/media/publications/downloads/1924_PDF.pdf [Accessed 18 Aug. 2018].

CHAPTER 7

LIMITATIONS OF AGRICULTURAL ENTREPRENEURSHIP

CONTENTS

Almost all the countries have a changing business environment, especially when it comes to Agricultural business where the competition is high in the domestic market. Various people get into the agricultural business and desire to become an Agricultural entrepreneur. However, there are various added limitations of agricultural entrepreneurship, which have been discussed in the chapter below.

The chapter highlights several issues that are related to policies formulated for agricultural entrepreneurship. It also deals with those factors that influence agricultural entrepreneurship. Various recommendations and solutions have been highlighted to overcome all these issues and barriers to achieve the organizational goals of an agricultural business.

7.1. INTRODUCTION

One of the most important economic sectors in the agricultural sector and it contributes with an important percentage of GDP.[1] The thriving sector in several developed countries is agriculture as it results in growth and development of rest other sectors in order to turn the wheels of the economy.

However, the sector is undergoing several challenges[2]. In order to achieve the goal of continuous growth, dealing with challenges and problems, association with globalization and also attaining sustainable development, agricultural sector initiates in competitive markets with suitable supportive and political procedures, in line with the growth of a commercial and empowered agricultural sector[3].

Entrepreneurship and entrepreneurs in different countries are given special attention. Entrepreneurship is like a practical and an objective concept that is actualized by the commencement of human life on Earth and to earn money and fulfill the life needs.

However, this challenge as a scientific notion has a little background. The first proposal of entrepreneurship came by Richard Cantillon in the late 1730's as one of the important factors engaged in the development of economic activities. Subsequently, Joseph Schumpeter, a foremost economist of the twentieth century, presented entrepreneurship as an engine of economic growth and studied its several features.

---

1 Morad Nejadi et al., 2007.

2 Ahmadi, 2009.

3 Morad Nejadi et al., 2007.

There are several definitions of entrepreneurship as provided by various scholars. Heinonen (2003) states that "entrepreneurship means the process of creating something valuable and different via allocating sufficient time and effort associated with financial, psychological, and social risk taking, as well as receiving financial rewards and personal satisfaction with the results obtained."

Usually, entrepreneurship can be measured as a vigorous procedure that comprises idea, alteration, transformation, and creativity. This procedure necessitates the application and use of energy and inspiration of people to develop and execute new perspectives and additionally practical solutions. In terms of social dimension, the entrepreneurial procedure is a set of actions necessary to recognize and measure chances, business, and operational recognitions, definitions, and acceptance of resources[4].

Entrepreneurship should be taken into account as one of the necessities of the twenty-first century, which is so-called information age, with concerns such as innovative technologies, globalization, quick alterations in the exchange, the strong rivalry in the field of business, etc. In such an environment, individual, and business entrepreneurship can be a point in the growth and development and it is one of the most vital factors that are efficient in attaining the anticipated future[5]. Because of creation and expansion of new innovative technologies, the standing and role of entrepreneurs gradually develop and variations and changes in the field of entrepreneurship have influenced the entire world so that work and trade will be continuously influenced by this issue[6].

In point, entrepreneurship is an innovative and growing attitude of sustainable agricultural development and actors and agents, together with technical-agricultural engineering consulting services. These companies have determined strategic and tactical work to enable entrepreneurship development within the companies and likewise to attain sustainable entrepreneurial agriculture interaction with other actors within the system of agricultural knowledge, information, and technology that can be completed with different systems.

Therefore, the essential part of technical agricultural engineering consulting services companies in the continuance of the aims of the agricultural sector, development, and strengthening of entrepreneurship in

4 Wright, 2007.
5 Qanbar Ali, 2008.
6 Faraji, 2011.

these companies can enhance their performance. This can, therefore, result in growth and development in the agricultural sector. Clearly, personal profit is the main objective of all these entrepreneurial processes; however, it is trailed by generating suitable situations for entrepreneurship at the native, county, and national levels[7].

In the recent time, therefore, rural development specialists believe that with the intention of authorization of rural communities, decision making, and rural management. This should be developed within the local community and outside factors play a role of organizer.

However, in recent decades across the globe in several countries, the misconception of the growth procedure especially rural development resulted to the development of a big government, which ignores the contribution of individuals with creativity and innovation and their cooperation in the development procedure.

Hence, the growth of entrepreneurial activities in rural segments of the economy should be founded on the capacity and potentials of the villages so that to attain success[8]. Various studies have demonstrated that the most significant strategy for economic growth in cities and villages is dependent upon job creation and its most vital mechanism is entrepreneurship. Unemployment is reduced by entrepreneurship, it leads to an improvement in the productivity of individuals and resources, and finally, it improves the income of the society.

Agencies take into account entrepreneurship as the great dynamism of employment, politicians take it as an important plan to avert discontent in rural areas, and economists reflect it as a method to progress their income. Governments, policy makers and planners have assumed the part and rank of entrepreneurship as a source of growth in developing countries more than ever.

Over a period, governments are enthusiastically involved in motivating entrepreneurship. Taking into account the social, economic, and political structure of countries, entrepreneurship requirement is diverse as well. Hence, knowledge and information of the requirements are important for government programs, in this context; the important conditions for initiating entrepreneurs' activity, the governments will play a significant part in the development of entrepreneurship[9].

7 Nobakht, 2007.
8 Hosseini, 2009.
9 Obamuyi, 2008.

In the winter of 2010, the rate of youth unemployment in Qazvin (15–24 years) was 21%. However, in comparison to same season in 2009, the rate of unemployment in the province of Qazvin, focusing on employment and entrepreneurship for this particular type is important[10].

The comparatively high population of rural areas that has a conclusive role in the labor force and food production and supply causes the requirements to recognize the barriers and encounters of entrepreneurship in rural areas more than ever.

Correspondingly, taking into account the improvement in rural population and their level of education, entrepreneurship education must be appropriately executed in rural segments, and firstly, challenges, and obstacles to entrepreneurship in rural areas must be truly recognized[11]. As quoted by Ronning, 2007, "personal, environmental, and organizational factors influence the performance of entrepreneurship."

In 2005, Lord Kipanidze said that "the major limitations in the development of entrepreneurship in small industries in Bangladesh have been the governmental expenditure, trade policies, legal, and administrative problems and financial constraints." Most important factors engaged in the growth of entrepreneurship are infrastructure and environmental factors as taken from the perspective of Maya, 20022.

Also concluded by Hornzbay et al., (2003) that the association of actions, authority, and accountability, a division of labor, rules, and regulations, administrative support, structure, and hierarchy, organizational culture and access to resources are elaborated in the creation and growth of entrepreneurship in organizations[12]. In their education on the part of supports and incentives of management in Finland, motivating people, transparency of strategies towards the expansion of information connected to the growth of structural entrepreneurship have been established.

In another research, AlamBeigi (2010) discovered that three variables of suitable content delivered to the employee by the ICT organization, plans for generating database associated to activities of the organization, and provision of progressive technologies by the organization of the organization, have the maximum influence on the growth of entrepreneurship in organizations.

As believed by Ardakani and Taheri (2012), the obstacles to entrepreneurship in Iran are quite different from the obstacles delaying

10 Statistical Center of Iran, 2011.

11 Basiri, 2011

12 Histrich,2003

entrepreneurship in developing countries. The ineffective administrative law as an obstacle to entrepreneurship in the economy, lack of taxes and duties rules as the next obstacle, and lack of financial backing of young entrepreneurs is arranged in the third place of rank.

According to Yaghoubi et al., (2012), there is an important association between economic success and entrepreneurship. They are also identified as an important link between adherence to the fundamentals of entrepreneurs.

Jahangiri (2013), in his research titled "Explanation of the Pattern of Entrepreneurship in Public Organizations" established that clarification of the entrepreneurial model is appropriate for backing creative and entrepreneurial individuals in organizations. In the research of Ahmad Yaghoubi (2013), titled "Designing and Verifying the Pattern of Training Entrepreneur Managers in the Industry," it has been stated, "entrepreneur is driving force of economic development and the factor of creativity, innovation, and invention."

In a study of the factors affecting the growth of agro-entrepreneurship in rural segments, in the limits of Khodabandeh villages and in context to two groups of entrepreneurs and non-entrepreneurs by Eftekhari et al., (2013), segmented influenced factors into four groups and examining their effects in terms of every group.

From the perspective of farmers, correspondingly economic, environmental, infrastructural, social, and institutional factors are significant. Correspondingly, numerous researches on entrepreneurship have been shown in foreign countries.

In a study of Norwegian farmers by Ronning et al., (2012) decided that entrepreneurship development improves individual's access to information and lastly these information result in combined business networks and employment opportunities. In a study done by International Labor Organization (ILO) in 2013, it has been designated that direct job development has not been a positive investment since it not only causes job skills are paid less attention, but too it declines the motivation for using upcoming employment opportunities.

In the research of Loy (2013), it was decided that particular training of entrepreneurship rise skill of students and to grow entrepreneurship, if encouraged people are provided with suitable conditions and they attain essential opportunities and facilities, they can be an entrepreneur and have favorite effects for the organization.

7.2. ISSUES RELATED TO POLICIES

There are several issues which are generic, other, are shown by the below cases and are more suitable particular entry routes. Each case is diverse and an array of factors is obvious, however, the serious issues are dealt with these individuals are quite common such as availability of land, costs of capital, exit or transfer costs challenges. The role of policy, as showed by the following individual case studies, is contextualized in the short commentary in relation to every case.

7.2.1. Availability of Land

Irrespective of the low agricultural returns, the prices of lands remain high. New entrants have to fight with present farmers and landowners to obtain land. The arrangement of tight supply and strong demand has sustained to reinforce the value of farmland in Scotland (RICS, 2006).

In the past, the land was striking to adjacent farmers who expected extra output without considerable intensification in fixed costs (ADAS et al., 2004) and notwithstanding low agricultural returns (and the advent of the Single Farm Payment) in 2005, this seems to continue the case; farmer buyer activity remains to grow.

Farmers signify approximately 60% of Scottish farmland buyers, a higher proportion than elsewhere in the UK (RICS, 2006). Demand continues to be strong. In the first quarter of 2006, some of the main farmland price increases for two years were recorded in Scotland (ibid).

Non-agriculturists additionally have a noteworthy existence available in the market and the act of 'lotting' is currently boundless to provide food for this demand and pick up a higher cost for the vendor. The farm is broken into various little 'parcels' to attract to a variety of purchasers.

Farmhouses and houses are frequently sold independently to attract to those from urban regions seeking to live in the farmland.

The residential estimation of such properties (and the improvement estimation of conventional farm building within a commuting distance of population centers) exceeds that of the farmer running the property as a component of the customary agrarian business (Booth, 2005).

This too has suggestions for the trying new entrant. In a few zones, he or she is confronting solid rivalry from the farm sector for farmland and furthermore solid rivalry from non-farmer purchasers regarding the residential market.

Various industry reporters have communicated worry over the low take-up of new tenures (Limited Duration Tenancy (LDT's) and Short Limited Duration Tenancy (SLDT's,) especially the former) under ongoing Agricultural Holdings enactment (2003). Expecting to enhance the connection amongst proprietor and tenant, the Act proposed to free up the market to let land and energize new contestants into the business.

While the Act (2003) is seen to be a change on its predecessor (the 1991 Agricultural Holdings Act), the Act abolished the utilization of Limited Partnership occupancies and adequately refused the already most prevalent and normally settled upon multi-year tenure i.e. tenures are presently required to be for a settled term of either five (SLDT) or fifteen (LDT) years.

One method of thought is that the firmness managed by the legislation obstructs the take up of new style tenancies. The commitment on proprietors to prepare the holding to a sufficient standard is additionally viewed as a disincentive to let. Different reasons referred to incorporate, "vulnerability over the impacts of CAP Reform, tax suggestions for progression and proceeded with worry among proprietors or their operators in regards to the right to purchase" (NFUS and SAYFC, 2006:3).

"The arrangement for a pre-emptive right to purchase and the concern by a few landowners of the likelihood of an absolute right to purchase being presented in the future… seems to have detrimentally affected the praiseworthy destinations of the Act" (MacKenzie 2005:2).

Along these lines, while one of the principal points of the 2003 Act was to urge new participants to the business, the accessible confirmation proposes that less land is being offered for let and that the rented cultivate showcase has hindered (on the same page).

In addition, Mackenzie's examination presumed that the explanations behind the slow down incorporate, " the attraction of keeping land in hand and using contract farming agreements for inheritance tax relief purposes." Contract farming agreements tend to support bigger, built up agriculturists against which new entrants find it difficult to contend.

7.2.2. Capital and High Expenses Related to It

In Europe, the uncalled problem of an aging population that was involved in farming along with the hurdles faced by new entrants has got in to spotlight. A lot of measure and support have been extended by the government framework and policies and have been included under the earlier version of Development Regulations (RDR) (2000–2006), RDR during the time period

2007–2013 with the aim to promote the exit of existing and aged farmers (According to ARTICLE 22: Early Retirement) and assisting emergence of young farmers (ARTICLE 21: Young Farmers Schemes).

Until this date, there are many nations who have not promoted this kind of schemes and policies in their Rural Development Regulation and Programmed and intentions in this manner are not so clear now, without any further doubts, it can be stated that there are major resource management issues that require immediate attention, for example, Scotland.

As per the latest outcomes derived by consulting as it does not bode well to launch this type of schemes incoming future. Some of the people take it as a hurdle. NFUS and SAYFC (2006) have raised the questions on the current category of young farmers being released by the National Reserve who the new entrants were, in the stipulated tie period just prior to SFP, extending the concerns about the resourcing of the National Reserve.

Bothe NFUS and SAYFC, support the introduction of a new scheme of single premium for new and young entrants but as discussed previously, while installation is available under ARTICLE 21 of the draft RDR, no clear indications have been noted that Scotland will bring this measure into action.

7.2.3. Issues Related to the Transfer or Exit of any Entrepreneur

There are multiple factors that affect the entry to farming are directly proportional to the low exit rates. Due to this, farmers can now work to greater age/ some of the people enjoy the lifestyle, some of the people are emotionally attached to the farming and are excited about the criticality of ties[13].

The policy framework is such that it is attached to the pension provisions, but these provisions are inadequate due to which there is a lack of affordable housing for the farmers who have retired and there exists an inheritance dilemma, due to which farmers are reluctant to retire.

As per the regards inter-generational transfer and less visible factors, for example, the relation of farmers with their children, are also influential. For instance, the balance of power, inter-generation conflict, a lack of confidence in his successor's ability, the farmer's anxiety over his future security reservations about the changes his successor would make can impose heavily on the decision to exit.

13 Fennel, 1981

Succession planning is one of the emerging issues that has been ignored and received no attention in many businesses related to farming, yet a flawless financial transition taking place from older generation to the younger generation is deemed essential for the long-term sustenance of the business.

The viable enterprise is need of the hour for a new generation who are active on farms and the retirees must feel safe in terms of financial security. "Where all the capital is in the business, planning is made more difficult as there are few options" (National, 2006:1).

A quick dispense of large capital from the business in order to support the older generation put an extra burden on the next generation and makes the job more demanding from the outset. Early financial planning coupled with the "ability to accumulate off-farm income whilst maintaining a healthy farm enterprise dramatically improves the chances" (ibid) for the next generation, but "a lack of advice and assistance and poor financial returns" have hindered this process in many businesses (NFUS and SAYFC, 2006:4).

Young and active entrants are supposed to be more motivated and innovative in terms of longer goals and better able to adapt, this is general terms allow the sector to become more competitive, productive, and viable.

This point is brought into light in the Forward Strategy for Scottish Agriculture: Next Steps (2006:14) "Entrepreneurship and innovation are vital ingredients for successful diversification, so it is important to retain talented young people within the farming community and to attract enterprising new entrants." This point is understood and highlighted in Defra's industry-led, "Fresh start" initiative that also targets to motivate and encourage more and more young farmers into the business of farmers, by promoting the need to "attract skilled and committed newcomers"[14].

For many years, the problem of an aging farmer's population and the requirement to motivate younger farmers to be in the industry is one of the most important agenda of the European authorities. Areas of concern and arguments majorly discuss the concept of multifunctional agriculture that includes safe food, conservation of natural heritage, environmental, and animal protection, biodiversity, landscape preservation, rural tradition, and culture.

It has been discussed that this is not possible to achieve the multifunctional role of agriculture as demanded by consumers if the average age of retirement of farmers remains high. "The continuing reduction in the number of young

14 Defra 2004:1.

farmers leads to a weakening of the agricultural entrepreneurship in the sense of less capacity to face the opportunities of the market and to combine the functions of agriculture with the new necessity expressed by the consumers and citizens" (CEJA, 2003:1).

Till date, the family farm business remains the most traditional way to enter the farm business, but still, new young farmers have to face a lot of barriers be it the economic troubles and uncertainties that discourage prevent and hinder them from entering this business.

The longer the would-be farmer is elsewhere, the more difficult (in social and economic terms) it becomes for him or her to return to the home farm. With existing and potential farmers looking to leave the industry, the implications undoubtedly point towards a more concentrated sector – decreasing farm numbers with a corresponding increase in farm size.

Areas nearby to population centers will continue to witness polarization – increasingly concentrated, large-scale farming at one end of the spectrum and small, lifestyle farms at the other. Further alloreactivity and diversification will continue to play an important role in boosting farm household incomes in these areas.

7.3. FACTORS INFLUENCING AGRICULTURAL ENTREPRENEURSHIP

7.3.1. Individual, Personality, and Behavioral Characteristics

The foremost issue and one of the most important areas were factors that are related to personality, behavior and individual characteristics.

This factor was recognized on the basis of viewpoints of the lack of certain personal capabilities and features were the major problems coming in front of the development of women's entrepreneurship in rural areas. Hence, strengthening or creating the capabilities and features needed for rural women entrepreneurship can be presented as an influential and important factor in improving the rural women's entrepreneurship process (Figure 7.1).

Figure 7.1. Poor Farmers and poor facilities lead to poor performance.

Source: https://upload.wikimedia.org/wikipedia/commons/b/b0/Sierra_Leone_rice_farmer.jpg

In other words, it can be concluded that the present state is due to the weakness of the entrepreneurial environment in rural communities and the poor performance of educational institutions as well as, women's personality, individual, and behavioral traits. As per the results of content analysis of discrete factors influencing women's entrepreneurship development, the following notions can be noted as significant factors:

- A relatively low level of confidence among rural people;
- Poor ability to take risks and to foresee the future of farming in rural areas;
- No acceptance of women taking leadership roles by the public;
- Physical limitations in performing some activities;
- Tendency to be isolated and to withdrawal;
- Limitations in communications and others;
- Low levels of education and knowledge.

7.3.2. Family Characteristics and Conditions

Economic and family conditions of the family environment were recognized in the initial stages of the research and are one of the major factors influencing eh entrepreneurship development in rural women. The major factors that influence the above-mentioned problems are mentioned below as (Figure 7.2):

- Undesirable attitude of males towards women's economic and social activities.

- Biasing on the basis of gender in rural families.
- Minimal participation of women in decisions related to family and home affairs.
- Overloaded with duties of the family (child care, household activities, etc.).
- Male chauvinism towards the scope and type of women's work.
- Men's reluctance to support women in home affairs.
- Amount of economic independence of women in the family.
- Women's confines to link with different groups of society.
- Inability of the family to support women entrepreneurs.
- Patriarchal mindset in most of the families around the globe.
- Dependency of families on men's income.
- Family size (in particular the number of children).

Figure 7.2. Together a family can contribute a lot towards Agri-business of an entrepreneur.

Source: https://www.usda.gov/sites/default/files/styles/featured_image/public/usda_blog_images/8032_7929972148_d3297e545f.jpg?itok=SKpGsruf

From the results of this section, it can be said that most of the answers depict that conditions of families in rural areas in terms of socio-economic aspects have an adverse impact on rural women's entrepreneurship development.

7.3.3. Knowledge and Skills (Education, Experience, and Communication)

Managing entrepreneurial and starting up businesses by entrepreneurs requires a complete skill set of features and abilities that can be learned through empowerment and training.

It is evaluated that by applying knowledge and skills within the entrepreneurial education and vocational courses, individuals can learn the ability to manage entrepreneurial businesses and to make decisions consciously. The following general ideas were resulting in the condition skills and knowledge:

- Inadequate entry for rural women to the type of technical and vocational training provided in cities.
- It is not possible to continue complementary education.
- Partial access to various types of education.
- Less economic work experience.
- Less number of women's associations and institutions.
- Insufficient infrastructure and facilities, i.e., it in villages.

7.3.4. Social and Cultural Conditions (Society and Rural)

In concern to the approached of entrepreneurship, the development of a cultural environment that supports the entrepreneurial efforts are always considered a long-term aim in a society. But, the short-term goals such as increasing abilities, motivation willingness, and skills of people to set up entrepreneurial businesses and increase community support of risky businesses should be considered.

The socio-cultural traits, that include opinions, beliefs, social norms, and customs, can have a lasting impact on the entrepreneurial activities of women in rural areas and reduce the interest and willingness of rural women in venturing and economic activities. The most important factors and variables associated with cultural and social features are mentioned below:

- Undesirable attitude among rural communities in respect of women's businesses.
- Public's opinion about women's inabilities as compared to men.
- Societal belief and norms about women's role in various activities.
- Negative mentality of rural communities in respect of women's communications and contacts with other social networks and

groups.

- Biased value placed on women's economic activities by society.
- Male-dominated and patriarchal culture in the village.

7.3.5. Access to Facilities and Services (in the Startup, Management, and Development of the Businesses)

Although consideration to skills and training education and providing cultural and social conditions are important to entrepreneurship development activities in rural areas, in most cases, implementation of the entrepreneurs' plans requires equipment, facilities, various services, and resources. The most important barriers mentioned are given below.

- No access to bank credits and loans.
- Investors' unwillingness to invest in businesses by women.
- Familial inability to offer requirements to promote women's business.
- Low participation and cooperative spirit among rural women in financials.
- Insufficient services to supply rural women with marketing techniques and skills.
- Few possibilities for the supply and purchase of products in the market.
- Lack of women's business development centers and incubators.
- Lack of transportation facilities crucial for accessing facilities and resources.
- Inability to provide needed collateral for resources and facilities.
- Few personal assets and little capital of rural women (such as property).

7.3.6. Economic and Financial Conditions

The economic empowerment and the financial situation of rural women is another factor impacting the starting up of businesses and the promotion of the development process of entrepreneurship by rural women. The barriers are the following:

- Women's small initial capital for setting up businesses at the entrepreneurial level.

- Poor access to credits and bank loans for rural women.
- Capital owners' reluctance to invest in rural women's economic activities.
- Lack of financial and credit associations in rural areas.
- Male-chauvinism in the rural family economy.
- Inattention to women's economic plans.
- Limited assets and savings of rural women.
- Women's inability to mobilize financial facilities and resources at the village level (in the form of cooperatives and enterprises).
- Lack of insurance systems for women's economic ventures.

7.4. CHALLENGES FACED BY WOMEN AGRICULTURAL ENTREPRENEURS

The efforts made by women in the agricultural sector have actually revolutionized several production practices in the past few years. Women are growing strong with every passing day when it comes to agricultural entrepreneurship. However, there are places where women Agri-entrepreneurs are overlooked because there is a perception already made that women from rural areas are conservative and have no knowledge about agriculture and various tools and implements used for production practices.

Women have always stood forward when it comes to agriculture and have been an indispensable part of the agricultural labor force, constituting an average of 43% in developing countries (Figure 7.3).

Figure 7.3. Women and agriculture.

Source:https://upload.wikimedia.org/wikipedia/commons/7/7a/2DU_Kenya_86_%285367322642%29.jpg

However, these women Agri-entrepreneurs come across various challenges when they are up for having access to the agricultural market. There are various criticizing barriers stopping women to conquer the world of agricultural entrepreneurship. Some of those challenges or barriers are:

- **Access to land:** The women help in producing almost 70–80% of crops and have access to only 1% of the land. In many countries, especially developing countries women have no rights on the land their ancestors have, it goes directly to men, and this shows that somewhere or the other there is lack of security in ownership, and in most cases of the one percent that own smaller plots to men that are less fertile.

Providing access to land to women the same way men have is only going to help in increasing the production of several crops in developing countries. This increased level of production in the agricultural sector could rise by almost 20 to 30% if women get the chance to have access to the land for agricultural practices.

- **No access to funding:** Securing credit is one challenge that women generally come across when they want to get into entrepreneurship, as people do not show much thrust upon them. It is believed that women are not much aware with the ins and outs of borrowing from an institution, and without assistance and support, they would not be able to have proper access to the much-needed funding for the Agri-entrepreneurship.
- Only 7% of the agricultural extension services are provided to women and less than ten percent of the credit offered to small-scale farmers.
- **Limited access to new practices:** Most agricultural extension is concentrated towards large-scale commercial farming. In such cases, very little surveys and researches are conducted on techniques, which are related to small-scale farming. These small-scale farming are generally taken care of by women Agri-entrepreneurs. Therefore, they have limited access to new techniques and practices for the production of crops.
- **Limited access to technological advancements:** With the advancement in the agricultural sector, there has been an advancement in the techniques used for agricultural practices. Some of those practices which have been encouraged with technological advancements are; plowing, cultivations,

plantation of crops, harvesting, and irrigation equipment. Most of these advancements are aimed at a male-specific audience, with improvements aimed to accommodate their requirements.

Women farmers often lack the expertise and the confidence to use the improved technology and most of the new technologies.

- **Comparatively less market opportunities:** Other major barrier women Agri-entrepreneur come across while performing their activities towards the agriculture sector is that they have less market opportunities when compared with the opportunities provided to men. This is so because, women have lack of market research and limited information regarding various market leaders and consumer behavior and preferences, which leads to their confinement towards the local, markets where prices are generally lower than the prices prevalent in the urban markets.
- **Lack of infrastructure:** Women entrepreneur merging from the rural areas always encounter barriers from transportation ends and from logistics end. If women get the opportunity to have access to transportation and logistics facilities than this might help them in improving the mobility. This would allow them to get a large number of products in the market for sale and would get in touch with more customers on a regular basis in a given period.

For the developing countries, it is important to empower women and invest money and them in their efforts so that they are able to prove their capabilities in an open market scenario specifically in rural areas.

Figure 7.4. Women have agricultural knowledge but not permitted to channelize it.

Source: https://www.usda.gov/sites/default/files/styles/featured_image/public/usda_blog_images/3769_13136064025_bfe9e9ab32.jpg?itok=Izz8TcZE

This will also help in increasing the productivity of various crops, which would somewhere help the country's economy to grow at a significant level and would contribute towards eliminating hunger and malnutrition at a larger rate (Figure 7.4).

7.5. AGRICULTURE AND ENTREPRENEURSHIP: ANALYSIS OF THE BARRIERS

7.5.1. Higher Cost of Machinery and Tools

To achieve success in agricultural production, it is important to have a significant growth in the country' economy. This economic growth can be achieved if there are better tools, equipment, and machinery available for the production of agricultural crops and products. High prices of agricultural machines and equipment, with the fall of agricultural product prices, demand intensive usage on behalf of the farmers in order to use them in their full capacity.

If this does not happen systematically, then there were, chances that there would be an adverse impact on profit and cause losses as there would be high fixed expenses. The temporary position of most of the agricultural economies does not enable fast redirection to market production exclusively.

The higher cost of tools, implements, and machinery is one big factor, which acts as a barrier to agricultural entrepreneurship. This is also considered as one of the biggest issues in the case of the agricultural mechanization application. Expenses machinery and tools are not generally used in those areas, which do not produce crops at a large level.

However, there is always a solution to a problem and in this case; the solution could be that the entrepreneur could have a joint procurement of the expensive equipment by several homesteads or usage of other parties' services. Thus, for example, Pihtrager, and Wagner (2002) point out "association of agricultural economies can achieve more rational use of agricultural machinery and equipment, and become more competitive in the market."

Machine ring is one way where the neighbors have the opportunity to represent a self-organized form of cooperation, which covers a larger territory and has clear principles for work payment according to the well-known prices agreed in advance.

The machine ring system was first started in Germany almost 40 years ago, and later they spread across most of the countries in West, Central, and North Europe. The development of SMEs in agribusiness should be based, apart from their own sources, on the investment of significant funds through various forms of credit (Figure 7.5).

Figure 7.5. Agribot: An expensive tool used in producing agricultural products.

Source: https://robohub.org/wp-content/uploads/2012/11/Bonirob_2.jpg

7.5.2. Agriculture Finances

People who aim at becoming an agriculture entrepreneur always starts up with an SME and farmers and SMEs do not have access to large funds when compared to any other sector like, E-commerce in developing countries. The lack of funds provides a narrow scope of credit to such budding Agri-entrepreneurs forcing them to quit their passion.

Various organizations and institutions offer agriculture finances but are insufficient. Therefore, it has become important to bring changes in such organizations and the way people approach them for funding. Funds to such entrepreneurs can be provided through the institutional support and share capital from the banks, credit associations and leasing companies (Bogavac Cvetković et al., 2010).

There are various financial barriers to Agri-Entrepreneurship, like high-interest rates, higher mortgage costs and obligations, and other types of credit insurance. For any budding entrepreneur, it is not easy to overcome such barriers. Therefore, to have a successful business in terms of better quality and more production it is important to have an active and flexible price policy which is related to price parity, incentive mechanisms of tax, credit, and other policies with the simultaneous change of the social status, work, and life conditions for the farmers.

This illustrates that there should be an arrangement where all the urban along with the rural population are taken into consideration parallel to other non-agricultural population like the consumers and retailers. The strategy of agriculture development must be a part of the united development strategy of the complete economy and society. This is the only way to create the basis for the application of the long-term economic policy in accordance with other development policies, and to create the basis for their harmonized action.

7.5.3. Establishment of Educational Systems in Rural Areas

To have a well-established entrepreneurship or a business in rural communities especially for budding rural women entrepreneurs, it is first required to have a well-structured and concrete base for those people who have diversified social and cultural backgrounds through the establishment of various educational systems and centers.

For the construction of such educational systems in rural communities' various studies were conducted in which the authors have shown a strong relationship between the cultural attitudes of people in different communities and the type of entrepreneurial behavior. In addition, the same group of researchers found that it is important to determine the level of effective entrepreneurship and for its determination, cultural differences act as an important factor in different communities of rural areas.

Hence, it can be evaluated that, for having a systematic framework for Agri-entrepreneurship, it is essential to have a cultural environment within the communities as this would help the people to value entrepreneurship which would further help in improving and supporting entrepreneurship (especially women entrepreneur).

Lundstrom and Stevenson (2001) argued that, "improving the entrepreneurship culture is one the most important factors in developing entrepreneurship in communities." On this basis, one of the major purposes for entrepreneurship development programs for rural women can be to foster the entrepreneurial culture in rural communities and particularly in rural women.

In rural areas, it is important to provide education related to entrepreneurship as this would help people in many ways. For example, it will help in constructing a strong level of knowledge, attitudes, and skills amongst the people, especially for the women who desire to become an Agri-entrepreneur in the future. These education systems would help the rural people to overcome existing weaknesses relative to men and women.

Research conducted by the ILO (2006), Duening, and Sherrill (2006), Lee, and Wong (2005), Solomon et al. (2002), and Saber (2003) have emphasized entrepreneurship education as an essential element in entrepreneurial development programs.

According to the ILO (2006) study, "the lack of education and empowerment as well as finding the appropriate way to operate entrepreneurship education programs is the fundamental problems underlying entrepreneurship development programs." The entrepreneurship development programs for rural women, in addition to building culture and awareness, motivation, and education, must factor in a supportive atmosphere for entrepreneurship development.

7.5.4. Support From Public and Private Sectors

The Rural Development Council of America (2001) is in the support of introducing a kind of structure that would support entrepreneurship in various rural areas concerned with agriculture. These support structures would play an important role in representing rural entrepreneurship development and would help in formulating various policies to support them in several ways. Therefore, it important to gain both the public and private sectors' support for developing rural entrepreneurship in the agricultural sector.

Apart from these support structures, it is also essential to focus upon those institutional and organizational structures that might be capable of facilitating and accelerating the process of entrepreneurship by rural women. In this context, the main agenda behind having institutional structures is to achieve entrepreneurship development programs for rural women. These institutions would also help in determining various barriers to agricultural entrepreneurship and other constraints faced by the entrepreneurs during their entire journey.

Similarly, the International Labor Organization (ILO, 2006) has reported that, "administrative barriers to entrepreneurship and the enterprise development process, including lack of support for entrepreneurs by tax organizations, administrative problems, high costs of registration and start-up of enterprises, rules, and regulations used by banking services and financial resources, and lack of transparency and changes in administrative procedures and programs of entrepreneurs."

Within the agriculture community, there are people from diversified communities and therefore there are more chances of having complexities in a business scenario. In such cases, the entrepreneurs are required to adopt such practices that would help them in bringing cultural, social, economic, and educational systems together and in harmony with administrative and financial structures and other mechanisms.

At the same time, it is essential to keep control over various policies as this sector has multidimensional characteristics and requires planning for each dimension to achieve entrepreneurially. Therefore, involvement, and cooperation of various governmental and NGOs is essential to developing and planning the policies.

In fact, there are several issues that are required to be acknowledged during various policy-making criterion and it is required that a different angle should be adopted for a different approach. The International Labor Organization (2004) argues that "a national program for employment should match the macroeconomic policies and the policies of other sectors; otherwise, entrepreneurship is isolated."

From the past few years, there are various factors that have been identified for the purpose of analyzing these factors' influence over a business and other farm operations, as these factors are somewhere or the other responsible for the success of the business's or its failure.

There are various factors that are similar and are equally responsible for the success of an Agri-business and these are the same factors, which are responsible for the success of any other business (non-agriculture). However, little research has been done on the perceptions of young, would-be entrepreneurs and how those perceptions might differ between the agriculture and nonagricultural groups.

However, it is not possible to measure anybody's point of view; the results of the study performed on young entrepreneurs suggest certain points of similarities and differences in the two businesses, i.e. agricultural and non-agricultural businesses. When asked about perceived barriers-to-entry, it has been analyzed that there is a significant level of difference in the test-statistic between the small-business and small-farm groups in the areas of competition, land/locality, risk, and unstable prices. The group means in each of these areas were higher for the small-farm survey participants.

For the analysis purpose, a hypothesis was conducted where the test-statistics showed no significant difference in the perception of capital, experience/education, and interest/commitment between the two groups of businesses. Under this hypothesis, various surveys were conducted at different venues to collect data. The survey comprises of questions related to people's perceptions over improved prices for producers and retailers. No significant differences in means were found in most of these options, with the exception of municipal markets, farmers' markets, wholesale markets, and private contracts.

These four market options were found to have a test-statistic that was significant, with the small-farm group having larger mean values. According to Cardozo, 2012, "with many farmers and ranchers in Paraguay producing commodities, cash crops, and beef cattle, such differences in perceptions in market options between the two groups can be anticipated."

7.5.5. Risk Taking Factors and Personality of an Individual

Every individual who has a desire of becoming an entrepreneur should have a personality type and willingness to take the risk. These are also some of the important factors, which help in analyzing the potential of a business in the end. Research conducted by Caliendo and Kritikos, 2008 suggests that "different personalities tend to be more willing to try new ventures and take risks, while others seem more hesitant and cautious of the idea."

However, when the hypothesis was conducted for this kind of study then there was no significant difference between the two groups regarding their perception of the importance of a business owner's willingness to take the risk. A significant difference in the mean values was found in the perception of each group about personality types of successful business owners.

Lussier, 1995 suggests that "a number of studies have suggested that controlling costs and keeping fixed costs low are crucial for an enterprise's success." Therefore, whenever a businessperson or a farmer tends to grow his or her business, then they are required to purchase a large amount of assets on credit basis. The study by Lussier also analyzed that, "The perception of the level of correlation between the percentages of assets held in debt and business/farm success." This was found to be significantly different between the two groups.

A large portion of the small-business group perceived this factor as having a larger correlation. The same was true with two more perceptions of the level of correlation between diversified products and business structure and business/farm success. It has been evaluated that, all these factors are perceived to have more correlation by nonagricultural groups in comparison to agriculture groups. The size of the operation and the perception of repeat clients about their correlation to business success were also significantly different between the two groups.

Years of experience, the enterprise as one's primary occupation, the percentage of spouse involvement, and the adoption of new technologies, among others, were found to have no significant difference between the two groups.

7.6. CASE STUDY: BARRIERS OF AGRICULTURAL DEVELOPMENT IN IRAN: A CASE STUDY OF FARS PROVINCE

7.6.1. Introduction

According to the U.S agriculture census, agriculture is described as a place of three acres or more on which any field forage crops were harvested or vegetables were harvested for sales of $100 or more (Mills, 1984; Mwaijande, 2007). Agriculture can provide a way for improving the potential economic of rural communities.

Agriculture is the core of the export market and it is accountable for one fourth to one-half of (GDP) gross domestic product in developing countries (Karbasioun et al., 2008). About 25% of the Gross National Product, 33% of employment, 25% of non-oil exports and 80% of food requirements have been provided by the agricultural sector in Iran.

Nevertheless, there is some evidence that agricultural development has some barriers. For instance, about 30% of the forests located in the North of Iran were destroyed during the last two decades. Furthermore, large portions of pastures and grasslands were rendered unproductive because of overuse by the cattle of the nomadic population and farmers (Darvishi, 2003; Karbasioun et al., 2008). The salinity of soil and water resources is a serious threat in many parts of the country (Siadat et al., 1997).

The research of this study explores the policy constraints as institutional barriers within agriculture sectors. Roughly, one-third of Iran's total surface area is suited for farmland, but because of poor soil and lack of adequate water distribution in many areas, most of it is not under cultivation. Only 12% of the total land area is under cultivation (arable land, orchards, and vineyards) but less than one-third of the cultivated area is irrigated; the rest is devoted to dry farming. Some 92% of agro products depend on water (Wikipedia, 2010a).

7.6.2. Barriers Identified for Agricultural Development

Information for this study was gathered from rural residents through FGD. A qualitative analysis was undertaken to determine viewed the barriers of agriculture development. According to the collected baseline data, farming is the most common occupation in the 7 villages in Jahan Abad, Koushkak, Hassan Abbad, Deh e Murd, Khvajeh Jamali, Tashk, Dehzir, Chah-Ghaz, Ghah- Sorkh, there were overall 65 participants with an average of 57 years old. All participants; were males.

They were chosen because of their engagements in agriculture products. The questions were asked about to barriers of agriculture development. The findings showed that agriculture development in their villages is without any certain planning for rural development. Although the FGD respondent referred to various barriers in terms of agriculture in their villages but in this study, we refer to some common barriers, which have been discussed in the majority of FGD groups. The most barriers in terms of agricultural development were including:

- **Salinity of water:** The majority of FGD participants believed there is no suitable water for irrigation in their village. They believe that their villages have many lands, but they do not have enough water to irrigation of these lands. Another problem associated with the expansion of irrigated farming is the overdraft of groundwater. In support of a finding Sadat et al., (1997) also state that the sustainability of our agricultural production is highly dependent on the "health" of our soil and water resources. However, the future of these resources is highly threatened by salinization and eventual decertification (Siadat et al., 1997).
- **Lack of resources:** The most participants in FDG groups mentioned to this issue as main obstacles to agriculture development. Lack of credit resources also was another barrier to develop agriculture development in their area.
- **Lack of human resource development:** Most of the young villagers immigrated to the citied and so, however, there are many educated people from the rural area of Iran but they do not

have an apathy to engage in agricultural activities.

- **Lack of government support:** lack of government support to provide funding for a poor farmer to develop their activities. Focus groups often complained about the lack of agricultural organization support to provide adequate facilities for farmers.
- **Lack of agricultural organization capacity:** FGD respondents believed the lack of capacity of agriculture was behind the failure investment for their agriculture product. However, at the end of any discussion, they refer to the barriers of agriculture development through government policy as well as local organizations.
- **Lack of agricultural knowledge:** However, in agricultural modern the role of knowledge is important. However, regarding the discussion with famers, they stated that the government did not support them with new knowledge. Base on my observation the farmers had traditional knowledge in agriculture. With mention to above focus group discussion about barriers of agriculture development, I summarized these barriers into two groups: community barriers and organizational barriers.

7.7. CONCLUSION

This paper has identified the barriers to agriculture development. Lack of capable organizations and community resources were an important element contributing to limited agriculture development. They referred to government policy and lack of local organizational capacity as main barriers related agriculture development. Clearly, the described barriers may not be only specific to Fars province; some of them may also be considered as common general problems of agriculture development in other communities in Iran.

Based on the findings, it can be suggesting that rural empowerment can be a tool for agriculture development in Iran. The findings of this study can be useful for academics, researchers, and all stakeholders involved in designing, assessing or promoting agriculture projects, which are in any way associated with general development goals. An understanding of the existing barriers in agriculture provides basic information for setting a policy agenda to enhance agriculture.

REFERENCES

1. Aničić, J., Maksimovi, G., & Vukotić, S., (2017). *The Possibilities and Limitations of Entrepreneurship Development in Agriculture in Serbia.* [eBook] Available at: https://scindeks-clanci.ceon.rs/data/pdf/0352–3462/2017/0352–34621701171A.pdf [Accessed 15 Aug. 2018].
2. Aref, F., (2010). *Barriers of Agricultural Development in Iran: A Case study of Fars Province.* [online] Available at: https://www.researchgate.net/publication/281640691/download [Accessed 16 Aug. 2018].
3. Bharthvajan, R (2014). *Women Entrepreneurs & Problems of Women Entrepreneurs.* [online] Rroij.com. Available at: http://www.rroij.com/open-access/women-entrepreneurs--problems-of-women entrepreneurs-.php? Aid=48589 [Accessed 16 Aug. 2018].
4. Bizconnect.standardbank.co.za. (2014). *AgriBusiness | 6 Challenges Facing Women in Farming | BizConnect.* [online] Available at: https://bizconnect.standardbank.co.za/sector-news/agriculturearticles/6-challenges-facing-women-in-farming.aspx [Accessed 15 Aug. 2018].
5. Jangchi, K. S., Mahmoodi, S., & Mesbah, A., (2015). *Analysis of Barriers to Agricultural Entrepreneurship Development from the Perspective of Agricultural Entrepreneurs in Qazvin Province.* [eBook] Available at: https://www.textroad.com/pdf/JAEBS/J.%20Appl.%20Environ.%20Biol.%20Sci.,%205(12)47–55,%202015.pdf [Accessed 15 Aug. 2018].
6. Jensen, B., (2016). *Entrepreneurs' Perceived Factors of Success and Barriers-to-Entry for Small Business and Farm Operations in Rural Paraguay.* [online] Digitalcommons.usu.edu. Available at: https://digitalcommons.usu.edu/cgi/viewcontent.cgi?article=5766&context=etd [Accessed 15 Aug. 2018].
7. Movahedi, R., & Farani, A., (2012). Analysis of the barriers and limitations for the development of rural women's entrepreneurship. *International Journal of Entrepreneurship and Small Business, [online], 15*(4), p. 469. Available at: https://www.researchgate.net/publication/264822566_Analysis_of_the_barriers_and_limitations_for_the_development_of_rural_women's_entrepreneurship [Accessed 15 Aug. 2018].
8. *Promoting Agricultural Entrepreneurship Among Rural Youth*, (2014). [eBook] Available at: https://www.iyfnet.org/sites/default/files/library/GPYE_RuralEntrepreneurship.pdf [Accessed 16 Aug. 2018].

9. Sinha, D.(2016) *8 Problems Faced by Women Entrepreneurs in India – Explained!* [online] Your article library. Available at: http://www.yourarticlelibrary.com/entrepreneurship/8-problems-faced-by-women-entrepreneurs-in-india-explained/41097 [Accessed 16 Aug. 2018].

10. Williams, F. *Barriers Facing New Entrants to Farming – an Emphasis on Policy.* [eBook] Available at: http://ageconsearch.umn.edu/bitstream/46002/2/Work17%20Williams.pdf [Accessed 15 Aug. 2018].

11. Yaghoubib, J., Yazdanpanah, M., & Khayri, S., (2011). *Investigating Barriers to Enhance Entrepreneurship in Agricultural Higher Education From the Perspective of Graduate Students.* [eBook] Available at: https://core.ac.uk/download/pdf/82011753.pdf [Accessed 15 Aug. 2018].

CHAPTER 8

FUTURE SCOPE OF AGRICULTURAL ENTREPRENEURSHIP

CONTENTS

Agriculture is the basic source of income and livelihood in many parts of the world and the latest reports indicate that there has been a gradual decrease in the young people taking up agriculture as a profession. This is mostly attributed to the lesser economic returns this sector provides to the people. For the humans to survive, this sector holds a prime importance and cannot be ignored by any means. One thing that can help keep up with the field is entrepreneurship and it thus becomes important to know about the future prospects of entrepreneurship in agriculture. This chapter gives the readers the same knowledge.

8.1. INTRODUCTION

The alternating factors involving the social, financial, political, and cultural aspects across the whole world have made the options of the farmers and the countries, meant for their existence and sustainable durability, very crucial. This also involves the importance of making it sure that the changes that they make in their particular and specific economic arenas, turn out to be successful. There are various reasons for the rapid developments in the field of agriculture, which make this growth an important and significant one. They are essential so that:

- The nations and their farmers become self-reliant.
- The countries across the globe may be able to meet the security of its citizens in terms of food and nutrition.
- The income and wealth of the countries and various rural regions in it may be distributed in a justified manner.
- The focus and aim to reduce and eliminate the poverty across the globe may be achieved.
- The ultimate aim of improving the quality of people's lives across the world can be met.

The developments and the growth in agriculture have drastic effects on the other fields related to it, which may lead to the distribution and vast reach of benefits over the whole economic arena and include the maximum number of the people in its grasp.

Alex in 2011 stated that the rise of the economies with free market has grown into the formation of a new kind of entrepreneurship, termed as 'Agripreneurship' and a rise in the need of the individuals separately, for the obligation of carrying out their self-owned businesses. There are a requirement and scope for huge amounts of investments and a number of innovations in this field of agriculture.

The environment for the social entrepreneurship is developing rapidly throughout the world. As there is the acceptance of social entrepreneurship across all the sectors, the field of agriculture is getting involved in it too. There is a rise in the number of social entrepreneurs who are getting involved in improving the conditions of the farming sector. The concept of being new, there are not too many people who have invested in this kind of sector till now, though some have gone on to inspire a lot (Figure 8.1).

Figure 8.1. Agricultural landscape.

Source: https://cdn.pixabay.com/photo/2018/01/19/09/26/field–3092043_960_720.jpg

8.1.1. Entrepreneurial Environment for Farmers

The entrepreneurial farmers work in a complicated atmosphere that is generally unbalanced and full of dynamism. They form a portion of a bigger group of individuals that involve various farmers, suppliers, traders, transporters, processors, and other stakeholders.

Every person taking part in this chain loop a significant role to play in the making of the items and getting them available in the market with the use of this chain. Every person in the chain is required to be an entrepreneur and make combined efforts to improve the working of the whole system and make better profits from their businesses. Entrepreneurship forms an integral part for the existence and sustenance of farming in a dynamic and extremely complicated economic environment across the globe. The entrepreneurial farmers view their fields as an enterprise that is meant for making good profits. They are extremely motivated by their desire to get good returns from their farm businesses and this enables them to take estimated risks for making profits and to develop their businesses.

Quite a lot of them suppose that the future for farmers is very limited and can improve only if they change their mindsets on the working of their farms into entrepreneurial ones. They should convert their thinking processes into a more positive outlook so that they may be able to increase the output for markets and earn good profits.

This idea of turning entrepreneurial may turn out to be extremely challenging for the farmers that have small farms. They may require assistance from other workers and established institutions to turn themselves into a better position as an agri-entrepreneur.

8.1.2. Entrepreneurship Dynamics

To grow into established entrepreneurial farmers, the concerned people need to have good technical knowledge, have an innovative mindset and a good foresightedness in order to plan their agricultural businesses along the right path by following all the stages of business development starting from the foundation and going through its survival and growth to achieve the targeted maturity.

The difficulties that these people encounter on their way may be many ranging from social hurdles, economic barriers, policies, and rules, availability of information and financial assistance and their own capability of managing and getting along with the risks and the opportunities to be utilized. Entrepreneurship is focused towards finding ways and methods to form and grow an agricultural business that may be hugely profitable. The words entrepreneurship and agripreneurship are quite widely known in the educational arena and the small businesses that make a portion of the agriculture sector. Dollinger has put agricultural entrepreneurship forward as the formation of advanced economic firms aimed at the developments and profits by taking up calculated risks and certain decisions in an uncertain environment in agriculture.

8.1.3. The Need for Agripreneurship Development

The requirement of entrepreneurship in agriculture arises from the demand for increased productivity and more benefits. It also arises from the fact that the people across the globe are facing major problems in the form of unemployment and poverty in economically backward areas.

However, following the formation of new economic reforms in 1995, like the acceptance of privatization, globalization, and liberalization and taking into consideration the aims and objectives laid down by the World Trade Organization, the projections are that the rural regions may progress at the same rate as the urban ones.

Entrepreneurship may be able to provide a lot of benefits in the economic terms in the agriculture sector which may be listed as:

- Increased and improved productivity;
- Formation of new business enterprises;
- Generation of new employment;
- Formation of innovative amenities and items;
- Development of the rural regions; and
- Increased financial output.

The agricultural entrepreneur is an energetic business manager who carries out different tasks that may be related to agriculture, making use of several resources such as physical resources, financial resources, human resources, and information, for the purpose of accomplishing a particular and specific set of goals.

8.1.4. Challenges

Farmers should develop some particular qualities to turn themselves into successful entrepreneurs. They require to have a thoughtful mindset and be ready to take up the challenges in making up their businesses as long-term enterprises with an aim to attract sustainability from it. They should know how to take up the opportunities and use them appropriately for their profits.

Some farmers that have small farm businesses do possess these qualities, but their methods of agricultural practices are still conventional. The productivity of the farmers currently is based on the quantity they need and not on the quantity that may be possible to be produced.

The need for them to develop an entrepreneurial mindset is due to the possibility of various risks that they might have to face in the future. A farm must have the required passion and be quite careful while making various decisions regarding their agriculture business with respect to the value chain that has its impactful effects on the profits generated from the agricultural business.

The other difficulties that align may be listed as:

- Market-related risk;
- No easy access to finance and credit;
- Lack of information;
- Low bargaining power;
- Vulnerability to economic shocks;
- Lack of proper training.

8.1.5. Opportunities

Farmers may be given a specialized training to run their farms as a profit-making business and to invest profits back into the business to generate long-term growth. They also must be encouraged and motivated to develop and adapt new technologies and share them with other farmers. Extension workers can facilitate partnerships among farmers, between farmers and researchers to work together to identify, develop, and test new technologies and practices to improve productivity and profitability.

They must be aware of the fact that they have greater opportunities that allow them to produce beyond just surviving. However, by changing their resource mix and overcoming access and risk issues, opportunities can be expanded. They should expand their survival farming to include some economic activities and can move towards the path of developing profit-driven farming businesses (David Kahan, 2012).

8.2. INVESTING IN AGRICULTURE

Below are six ideas for ways people could invest in agriculture. These are by no means the only six ways.

8.2.1. Crowdfunding

Chris Rawley of Harvest Returns to discuss their how their crowdfunding platform works. Essentially, it allows investors with as little as $5,000 to take an equity stake in a farming operation. Harvest Returns primarily works with accredited investors, but there are opportunities for the rest of us to get involved. Crowdfunding sites are becoming more popular and will likely continue to provide opportunities for us to take part in agricultural investing as well as diversify an investment portfolio across geographies and commodities.

8.2.2. Peer-to-Peer Lending

Peer-to-peer lending allows investors to lend money in small amounts. The big differences are that the lender is not taking any ownership of the property or venture, but instead collecting interest on the principal.

Companies like Kiva allow you to lend to farmers in developing countries, while platforms like Lending Club and Prosper allow for more generic (not age-specific) lending. These platforms also offer some great data on risk and expected the return.

8.2.3. Publicly Traded Agribusinesses

This is the most liquid of all the options, meaning the stakeholders are the most likely to get your money back out when they want it. Several agribusinesses have reached the point where they have "gone public" with an initial public offering (IPO) of their stock.

Shares of large corporate agribusinesses such as John Deere (DE) and Archer Daniels Midland (ADM) are available for purchase, as well as smaller companies such as Marrone Bio Innovations (MBII) and Calavo Growers (CVGW).

8.2.4. Farmland REITs

A "REIT" is a Real Estate Investment Trust. This company owns, operates or finances income-producing real estate (Investopedia). In order to qualify as a REIT, the company must comply with several regulations, including paying back 90% of its taxable income to shareholders.

For a very modest amount (less than $10 per share in some cases), a person can own a small piece of a very large farmland holding company. Usually, but not always, these REITs make money by renting out the land. A couple of the larger publicly traded farmland REITs is Farmland Partners (FPI) and Gladstone Land (LAND).

8.2.5. Commodities

As with all of these, it is advisable to take time to learn the risks and the market dynamics before putting real money on the line. That said, this is one way that one can "store" their money in an asset that will theoretically benefit if demand for agricultural products outpaces supply. One can also look at commodity funds such as PowerShares DB Commodity Index Fund (DBC) that track all commodities.

8.2.6. Start Farming

While perhaps not a popular opinion these days, it is still believed that it is possible to get into farming, even without a million dollars. In fact, farmers that are making it happen multiple first generations.

This is by far the HARDEST and MOST TIME-CONSUMING route that will require significant sacrifice, but it is possible. Those that are making it happen are often finding niches, developing relationships with long-time farmers and landowners, working long hours, and making great sacrifices. However, it is imperative that their hard work will pay off and they will continue to acquire equity and assets in their agricultural ventures.

8.3. THE ROLE OF AGRI-ENTREPRENEURSHIP AND FARMING INNOVATIONS FOR YOUTH IN AGRICULTURE ENGAGEMENT

The conference titled, " The future of our food: What role can agri-entrepreneurship and farming innovations play to engage young women and men in the agricultural sector?" was facilitated by the Center for Development Innovation (CDI) of Wageningen University and Research Center. The conference and its members examined a wide scope of limitations, difficulties, and openings that depended on solid encounters of associations.

Figure 8.2. African agriculture.

Source: https://upload.wikimedia.org/wikipedia/commons/7/7a/2DU_Kenya_86_%285367322642%29.jpg

The point of the morning session was to rouse and gain from each other to enhance the commitment of youth in agriculture. This session followed up on the December 9, 2015, workshop titled, " The future of our food: How to engage young women and men in the agricultural sector?" CDI, AgriProFocus, the Royal Tropical Institute (KIT) and Young Professionals for Agricultural Development (YPARD) facilitated this workshop at the close of the year. The Food and Business Knowledge Platform gave a briefing on this seminar held in the month of December in an issue in January 2016 (Figure 8.2).

8.3.1. The Capability of Persistence and Youthful Sheets Inside Cooperatives

Zayaan Khan of the Slow Food Network in South Africa commented on the significance of grassroots activities for youth in the field of agriculture. The entrepreneurs working small-scale food industries in South Africa demonstrate the significance of persistence as the systems in which they take part, consumes a lot of time to develop itself.

By beginning at the regional level, systems can be constructed and exercises can be organized, at the same time as the societies and ethnicities are being resurrected. These kinds of activities are expected to take youth back to a profession in agriculture.

Furthermore, Marije Clever told her story which focused on her being a farmer for the dairy organization Friesland Campina and alongside other youthful agriculturists, they both endeavored to impact the basic leadership procedures of the firm.

Despite the fact that they do not have the authority to give opinions, they have been effective and powerful in different issues because of their group of youthful leader individuals. Involving youth in the basic leadership process inside the value chain may be a strategy to connect with the young people throughout the world, to make them return to agriculture.

8.3.2. Slander of youth

With the primary introductions as a premise, five topics were additionally explained in the World Café dialogs. Simon Bailey of Aflatoun, a network association for the rights of children, tended to psychological requirements for the commitment of youth in agriculture and their entrance to value chains.

For instance, Aflatoun is battling with the difficulties of various leveled structures inside networks that keep youth from responsibility for or incorporation in basic leadership forms, especially in regional projects in Ghana and Uganda. Additionally, entrepreneurial mentalities, being free from criticism and mental prosperity are imperative necessities for fruitful commitment in agriculture. Several scenarios show the trashing of agriculture, including the economically backward peripheries that are related with destitution, the sentiment of disappointment and being abandoned.

On the contrary end, urban areas are viewed as wellsprings of riches, opportunities, and achievements. Institutional game plans, for example, directed financing or safety nets, can make the picture of agriculture as more appealing, particularly with particular projects that are pointed towards evolving states of mind, pictures, and intellectual limits that, most probably, could have some kind of effect, enough to make the difference. This approach has a twofold advantage:

- It builds the benefits that the young have and;
- Encourages them to enhance their mental and psychological status.
- It is likewise feasible, as it tries to reconnect diverse ages and exchange learning between them.

8.3.3. ICT as a Ripe Opportunity

In another session, the subject concentrated on how ICT can offer open doors for work created in the enhancement of agri-chains. Encounters have demonstrated that limit building, and preparation and education of youthful and potential entrepreneurs pay off. In the session conducted at the cafe, CTA, and others shared their fruitful exercises of drawing in youth in agri-business by utilizing ICT.

Sharing data about the cultivating business through online networking sites and applications, like Facebook and WhatsApp, has enhanced the abilities of young agriculturists.

Likewise, training them how to trade learning by means of online journals, for example, Yobloco, and web-based social networking programs, like Makuli MaYoung, has been emphatically adopted by young experts.

Moreover, assisting these farmers to get their voices heard at hackathons or at brooding ICT focuses has turned into an advantageous system (Figure 8.3).

Figure 8.3. Wireless sensing technology in Kenya.

Source: https://upload.wikimedia.org/wikipedia/commons/f/f4/CSIRO_ScienceImage_3719_CSIROs_Fleck_wireless_sensor_network_technology.jpg

The Ugandan digital space for monetary consideration Ensibuuko is a case of a youthful social venture that grew from such a program. It has turned into a productive organization that makes digital spaces for the answers of the various problems, for making money related services effectively available to the individuals who do not have access to banking and are not well served.

Further, M-cultivate Kenya is a digital platform that gives cutting-edge market prices through an instant message that is sent straight to farmers. It likewise gives clients the chance to interface with purchasers and ranchers in their regional spaces to offer their harvest. Through the social endeavor Hello tractor in Nigeria, the agriculturists having low income can gauge the quality of their soil with a cell phone. Such groundbreaking ICT instruments and their wide extension, in addition to the way that they are being produced by young experts, ought to be given more consideration.

8.3.4. Comprehensive Arrangement on Agroecology or Agribusiness: The Significance of Systems

At the cafe assembly on the subject of policy instruments, individuals from India, Kenya, the Philippines, Uganda, and Indonesia traded their procedures on what is working in their nations to draw youngsters to agriculture. Members discussed the requirement for constructing and having trust among new and young entrepreneurs. It was deduced at last that young entrepreneurship can just be created in the privileged and empowering situations, ideally in different enterprises that are developed together publicly and privately, that permit component of mindfulness, access to assets and improvement of skills.

At another different session, the idea of agroecology, that is, the harmony between environmental soundness, financial feasibility, and social equity, was at the focal point of the discussion. On one side, it was contended that agroecology enhances agriculture's profitability and manageability.

In their contention, teacher Irene Cardoso of soil science at the University of Viçosa in Brazil indicated the report by UN Human Rights Council' Special Rapporteur de Schutter that affirmed the capability of agroecology. On the other hand, the suitability of agroecology was addressed from numerous points of views like:

- How is it going to bolster an undeniably increasing global population?
- Is agroecology extremely ecologically suitable, regardless of whether subjects such as soil composition may be considered actively?
- How is it going to be acknowledged in an environment of imbalance throughout the world?

With everything considered, there was sufficient knowledge to think about the idea. There was additionally a plan to not negate but rather consolidate all the best parts of this idea. The discussion with the Slow Food Youth Movement YPARD later inferred that networks and decent variety are vital in drawing in youngsters to agriculture. It could prompt imaginative arrangements and flexibility on a local level.

8.4. FACTORS THAT INFLUENCING YOUTH INTEREST IN AGRICULTURAL ENTREPRENEUR

Albeit numerous individuals know about the critical benefits of the agricultural segment, despite everything, it requires a stupendous effort to pull in the young people to take up agriculture entrepreneurs. Along these lines, perceiving the elements that affect them to end up as entrepreneurs are critical as it can be utilized as a direction to strategize and to advance agricultural division.

A few factors affect the young people to grow up into agricultural entrepreneurs. These components, which are drawn from the audit of past researches on the content available on the topics, are:

- Socio-demography;
- Attitudes;
- Acceptance;
- Knowledge.

8.4.1. Socio-Demography

The Socio-Demography is the principal factor that influences the demeanor and acknowledgment of youth towards agriculture entrepreneurship. It incorporates gender, age, pay, area, and culture. According to Silva and others from 2010, demography factors are the elements that affect mentality towards contract farming. Demographic factors that had been examined were gender, age, salary, region, and culture.

Gender is one of the markers for the elements that assume a part in deciding the demeanor and acknowledgment of youth towards entrepreneurship. Anyway, according to Konavalchuk from 2008, gender did not significantly affect work execution. However, in a research by Gidarakau in 1999, it was concluded that women were found to have a negative approach towards contract agriculture in contrast to men.

Age is intended to indicate the individual age suitable for farming exercises. There were noteworthy contrasts between age bunches concerning approaches towards contract agriculture according to Man from 2007.

Uli and others (2010) expressed that the study conducted by Salleh and Azril (2009) and Ezhar and others (2007) uncovered that the normal age of agriculturists in Malaysia is surpassing 46 years of age. These discoveries convey to the people one point, which is that Malaysia needs a reinforcement agriculture network comprising of youth that involves both the male and the female members to assist the present group of ranchers.

Obviously, there ought to be activities to pull in the young enthusiasm towards agriculture. For the larger part of the general population, the most imperative factor in picking occupations is the compensation, i.e. the wages or pay. Young people had a belief that farming industry is anything but a lively industry as it produces just pitiful compensations.

Subsequently, according to Girdarakou from 1996, the mentality of young people mainly is that being engaged with agriculture area is viewed as a momentary affair, satisfactory as a response to the issue of joblessness, just for the time until a better arrangement can be made. Silva and others (2010) expressed that past investigation by Stephenson and Lev (2004) found that pay is certifiably not a noteworthy determinant in making an inspirational demeanor towards agricultural initiatives while a research done by Hyttia and Kola in 2006, found an entirely alternate view.

Largely, the individuals with low income tend to pick agriculture as their primary money producing work or as a side pay.

Silva and others (2009) stated depending on the fact that Kumar (2007) in their research noticed that the ability that contract agriculture has, to offer to the social groups, increases the possibility to develop youth acceptance of itself.

The research was done by Kumar additionally fortifies the exploration finished by Mann and Kogl (2003) which underscored that greater benefits gathered through contract farming, will be an impetus for having more individuals to have an acceptance for contract farming.

The prospects in agriculture can be huge and overpowering. Obviously, the mandate for food will never die down in spite of any kind of monetary circumstance in any given nation. Subsequently, the prospects for agriculture entrepreneurship will be boundless. The issues regarding financial backwardness and joblessness in the society may too be settled.

The poor people picked agriculture as one of their principal cash producing alternatives because of their faith in the capacity of agriculture in delivering higher efficiency with lesser investment. The individuals belonging to rural areas are constantly connected with agriculture practices including contract farming. In any case, it was discovered that no noteworthy distinction existed between those with an agricultural context and those with none of it.

8.4.2. Attitudes

Attitude is another factor that affects youth to be engaged with agriculture entrepreneur. According to Bahaman and others from their research in 2010, these dispositions can be shaped in view of an individual's level of appreciation or aversion on something. Generally, state of mind depicts either constructive or adverse perspectives of a man, place, thing or an occasion.

Attitude is an essential determinant of an individual's accomplishment in entrepreneurship. Kuratko and Hodgetts in 2004 stated that according to Robinson, the attitudinal approach is the ideal approach to clarify the entrepreneurial procedure since it interfaces the individual and the ever-changing attitudinal protest. A genuine and uplifting attitude is expected to help a man in picking and taking an interest in entrepreneurship. This is because the choice to give up a present vocation or agreeable life as it is anything but a simple choice to make, as entrepreneurship requests penances.

Young people being the eventual fate of the country must be delicate and sensible in their aptitude advancement. They should think for their future as well as of their families, social orders and to a bigger picture is that of their country.

Professions are critical for each individual since they constitute an assortment of advantages and capacities that have been covering an individual's life. Professions in entrepreneurship are becoming because of the way that they can offer unlimited opportunities.

Like some other abilities, entrepreneurship has its own remarkable level of advancement. Youth's temperament in choosing a profession affects the joblessness issues.

It is recognized that the quantity of youth constitutes 42% of Malaysian subjects and the present rate of unemployment among youth in Malaysia is 4%. In another insight, from an aggregate of 27 million Malaysian recorded, 41% or 11.1 million are young people and the expansion in the quantity of young people has increased the level of unemployment, particularly among college graduates. The high labor among youth can give the potential feeder to nation change programs towards the formation of agriculture entrepreneurship.

8.4.3. Acceptance

Acceptance is another essential factor in affecting the young people to grow into agriculture entrepreneurs. Young people will be even more sincerely arranged and have uplifting standpoint towards agriculture as a whole, and all the more critically, propelled to set up their own particular agriculture-related business. Acceptance as an articulation or suggestion by activity that exhibits consent to the terms and state of an offer, this ends up in the arrangement of restricting contract.

Offer through some demonstration of execution, show the level of intensity. This demonstration of practicing the power for something is offered without anyone else's input or another as expected by holding the subject of the offer. Contract farming offers to the network groups the possibility to expand youth acceptance of itself. The researches led by Kumar additionally fortifies the exploration finished by Mann and Kogl in 2003, which underlined that greater benefits collected through contract farming will be the impetus for having more individuals to have an uplifting attitude and acknowledge contract farming.

Along these lines, the presence of acceptance factor will, in the end, pull in more individuals to include in the agriculture segment and give this part the genuinely necessary uplift.

8.4.4. Knowledge

Knowledge is the fourth factor and considered as a vital factor to set up an enthusiasm for youth in agriculture entrepreneurship. Learning can prompt more aptitudes to youth as a planning to build up their own business. Othman and others (2010) featured that in the Seventh Malaysia Plan 1996–2000, the push for youth advancement was on creating aptitudes important for them to contribute towards the viable country building.

It was additionally specified that both in the Eighth Malaysia Plan (2001–2005) and the Ninth Malaysia Plan (2006–2010), the accentuation was to set up the young people with uplifting attitude, learning, and aptitudes in order to empower them to deal with the difficulties and financial changes that were realized by globalization (Malaysia, 2001: 2005).

According to Othman and others, entrepreneurship knowledge and education are vital for financial advancement, especially in enhancing the quality and expanding the number of future entrepreneurs.

Entrepreneurship training at the higher organization of learning is likewise growing in the types of scholarly and down to earth perspectives. There are certain connections between entrepreneurial action and financial results, for example, innovation, and monetary development. As it were, a recognition that is more positive may result if the rural education learning level of nationals were to be upgraded.

Knowledge too has turned into a key factor in affecting insight, and this pattern is set to be increased. In the 21st century, Knowledge gathering and application will drive peoples' recognition. At the point when this happens, it expands the likelihood of having more positive discernment towards contract farming. There is most likely that new logical knowledge combined with worries for the social, monetary, and natural parts of agriculture brought Malaysian agriculture division into a physical change.

8.5. FUTURE-FOCUSED ENTREPRENEURSHIP: THREE MEGA-TRENDS INFLUENCING BUSINESS IN RURAL AREAS

What does the eventual fate of business resemble, have an aftertaste like and feel like? While numerous organizations are attempting to stay aware of the fast rate of progress, future-centered entrepreneurs are reproducing the business scene mixing innovation with the human experience.

It can be hard to foresee the future as the world keeps on advancing at an inexorably fast pace; be that as it may, entrepreneurs and entrepreneurs in rustic spots can utilize a future-centered administration approach, which incorporates inspecting megatrends, to anticipate what's to come. Megatrends are worldwide movements that affect society, the economy and nature. The three megatrends molding the eventual fate of business that is going to be addressed are:

- The Rise of the Gig-Economy;
- Living with Purpose and Meaning; and
- The Decentralized Marketplace.

8.5.1. Megatrend 1: The Rise of the Gig Economy

The rise of the entrepreneurial individual and what is at present is referred to as the Gig Economy, will make new tasks and prospects for entrepreneurs, organizations, and networks. A rising pattern, the ascent of worldwide unemployment (International Labor Office, 2014), is supporting the development of the Gig Economy while making a requirement for the displaying of abilities and aptitudes in new and significant ways.

The capacity to offer, buy, and advanced products and ventures and in addition singular brands on the World Wide Web, has made a requirement for entrepreneurs to increase the value of the lives of others in inventive ways.

Individuals are progressively characterizing accomplishment for themselves, which does not generally incorporate a professional education and a nine to five work. From multiple points of view, it is currently significantly cooler to be a fruitful start-up entrepreneur, well-known blogger or YouTube sensation than a man working in an all-day work is.

According to Edelman Berland's study in 2014, roughly 53 million Americans are occupied with some way of outsourcing, which is adding around $715 billion to the economy. Further, the quantity of huge managers is anticipated to contract later on making individuals find new and creative approaches to bring home the bacon.

Independent work has been recognized as an indispensable financial apparatus for rustic individuals and networks (Goetz and Rupasingha, 2014). The rising Gig Economy might be one megatrend that can profit from provincial spots if innovation can be utilized to achieve both local and experienced crowds, inspired by obtaining products and ventures from outsourcing entrepreneurs.

Rustic people group and business pioneers ought to consider building their brands and in addition those of their representatives and local organizations. The ascent of interpersonal organizations and innovation has empowered individuals to live, purchase, and offer at whenever from wherever.

These same devices have helped individuals make their own particular organizations and plan their lives in ways that best suit them, which underpins the Living with Purpose and Meaning megatrend. Deliberate living is forming everything from medicinal services to instruction.

8.5.2. Megatrend 2: Living with Purpose and Meaning

Our public is rapidly advancing into a period portrayed by shared reason and importance. Pioneers must make an air of manageable development intended to ceaselessly advance associations while interfacing with workers and demographics to significant encounters and their meanings of individual satisfaction.

Individual satisfaction is a developing pattern in business and entrepreneurship. Individuals are hoping to diminish their levels of pressure and reinforce their sound propensities while finding more noteworthy satisfaction.

Moderation is one of the patterns molding the Living with Purpose and Meaning megatrend on the grounds that individuals are looking for not so much stuff but rather more life. Pattern setting places like Silicon Valley have grasped moderation, and this way to deal with living is affecting everything from living spaces, to work environments and website architecture.

Innovation may help future-centered entrepreneurs comprehensively add more reason and importance to the lives of the two students and workers. For instance, a gauge via Carbone and Nauth (2012) portrays the development of the "cloud." They talk about the cloud advancing from a place to store knowledge to a canny asset that will effectively enable individuals to enhance their lives.

Virtual specialists will move from being a mechanized type of telephone-based client administration to a customized type of help and help that gives data and even more vitally performs valuable assignments. For instance, such specialists may outline a week-after-week menu in light of a family's wellbeing profile, wellness objectives, and eating inclinations, and consequently, arrange fixings.

This conjecture gives a case of how entrepreneurs can utilize innovation to enable their associations to improve with an attention on what has to come. A change from cloud-based processing to cloud-based living and learning may enable pioneers to center around new benefit focuses, client encounters and workers in a more comprehensive way.

Two patterns supporting the Living with Purpose and Meaning megatrend include:

- Life progressively streaming together as opposed to being compartmentalized into work, family, entertainment, and so on; and
- The journey for more prominent prosperity, wellbeing, and imperativeness. Individuals are progressively learning and living in a hurry in both the virtual and physical faculties, which obscures the lines between family, self-care, work, and play.

How do these patterns affect country entrepreneurship?: As more individuals need to associate with their motivation, discover importance in their lives, and increment their imperativeness, organizations, and networks should adopt a more people-fixated strategy concentrated on significance, prosperity, and the mind-body association upheld by the correct interpersonal organizations, encounters, and advances. A couple of thoughts may include:

- Coordinating all-encompassing, sound living and self-mind standards into business rehearse with an end goal to make better worker and client encounters and to make more an incentive in items and administrations.
- Giving training to workers to enable them to reveal their motivation and make methodologies and structures to help important activities intended to enhance their professions and lives.

- Including physical movement and fun into the workday with an end goal to empower innovativeness, diminish pressure, assemble kinship and reinforce generally health (maybe an App or the astute cloud can help).

8.5.3. Megatrend 3: Decentralized Global Marketplace

Business is in the age of the "Engaged Consumer" who has a buffet of purchasing decisions. The enabled purchaser chooses what shopping encounters they need, when they need it and how they need to encounter it. Enormous Data is one apparatus that can be utilized to create showcasing efforts, think about human practices and decide designs.

It is likewise being utilized to look at and upgrade different zones of the shopper encounter. Rustic organizations can use Big Data to gain by buyer examples and requests while making more tweaked shopper encounters.

Knowledge is gathered with each snap, and a few measurements can be utilized very quickly. Retail internet business (orders by means of PC) and m-commerce (orders by means of cell phones) are anticipated to increment by twofold digit edges well into 2018 (marketer, 2014). Knowledge exhibits the proceeded with development of the advanced commercial center, and country organizations must discover approaches to benefit from these patterns.

The world is quickly turning into an "Add to Cart" society for everything without exception. People are making nearby, national, and worldwide networks of importance and change that are self-chosen and self-framed. A large number of these same people are changing trade and networks in new and creative ways. Innovation joined with the requirement for significance, self-articulation, and self-completion has made an inexorably decentralized commercial center.

8.5.4. Future-Focused Entrepreneurship: Cooperation, Creativity, and Purpose

Curiously enough, the most blazing employments at Facebook require a "human touch." Why? They are searching for individuals who can end deals in light of the fact that their income depends vigorously on advertisements (Anders, 2014). Maybe the most future-centered entrepreneurial pioneers are mixing "high-human touch" with "cutting edge" in new and intriguing ways that comprehensively increase the value of the lives of others.

Provincial entrepreneurs and pioneers with an attention on the future can utilize innovation to change the cycle of life. One individual cannot do much in a world that keeps on changing at an undeniably quick pace; rather, provincial pioneers and entrepreneurs ought to develop in networks with a soul of collaboration and inventiveness.

Together, with a typical reason, pioneers can look at megatrends, use rising advancements, and screen investigation while making significant encounters and enhancing country to put in efforts in was not yet envisioned.

8.6. FEEDING THE FUTURE OF AGRICULTURE WITH VERTICAL FARMING

On an average, the global food prices have increased by 2.6% annually in the past two studies. If this particular trend remains, it not only threatens a standard quality but also more disposable income goes in the direction of food, additionally, it threatens the overall food security.

The issues of hunger and malnutrition continue, particularly in developing countries. The problems of food scarcity have additionally been associated with violence and political unrest. According to the United Nations World Food Programme, "record-high food prices in 2008 prompted riots in 48 countries, including fragile states like Somalia and Yemen."

Increasing prices of food reveal fundamental trends resulting to let-downs with traditional agriculture. Vertical farming is known as a technology-driven model of agriculture. It may also offer a method to deal with farm output and food security in the years to come, even though it may not influence food prices in the upcoming time period (Figure 8.4).

Figure 8.4. Vertical farms.

Source: https://upload.wikimedia.org/wikipedia/commons/6/62/Vertical_farm2.jpg

8.6.1. Reasons for Conventional Farming Frustrating All

There are several requirements of field farming such as labor, amenable weather conditions, and sufficient sunshine for photosynthesis, irrigation, and frequently pesticides to protect crops. This has not transformed; however, it is easy to identify reasons of traditional farming no longer working together with it is utilized by using a framework people developed.

Though it may seem that the economies of the world are meaningfully influenced by unexpected events, the DRIVE agenda is founded on the concept that particular interrelated large-scale procedure, which determines the notion that particular unified large-scale processes, which determine the behavior of companies, administrations, and societies, also affect the future.

By examining demographic and social variations, resource shortage, differences, and instability, scale, and complexity, that can estimate the ways the future may unfold. Examined together, these megatrends can expose the root causes behind the shifts in traditional agriculture.

- **Demographic and social changes:** Food supply at the global level cannot maintain the increasing global population. According to the Food and Agriculture Organization of the United Nations, "food production must increase by 70% before the year 2050 in order to meet global food needs." It was also recommended that this development should occur against a headwind—urbanization is attaining arable land while concurrently pushing people away from farming as a profession.
- **Resource scarcity:** Agriculture consumes up to 70% of total global water consumption, in addition to its total cost. Given the assessment that half of the population of the world will experience a shortage of by 2030, the production method in agriculture is unsustainable. The inefficiencies of the supply chain further aggravate the scarcity effect. Unpreserved crops mark and spoil in the course of harvesting, packaging, processing, and distribution. According to a Natural Resources Defense Council report on food from field to fork to landfill, up to 40% of all crops are ultimately wasted.
- **Inequality:** Along with venerable issues of malnutrition and extensive poverty in developing countries, inequalities related to food prices have additionally come up in industrialized countries. In countries like the United States, the prices of fresh foods have resulted in susceptible populations to look for fat- and sugar-

laden treated foods with little nutritional value.

The results of these food "choices" is a global obesity epidemic together with an increase in diet-related diseases like diabetes. At the other end of the range, higher-income households are initiating the demand for additionally health-conscious "superfoods" like antioxidant-rich kale and protein-packed quinoa. As nationwide food needs and the prices of agriculture remain to increase, the potential of enhancing health and nutrition conditions are terrible for low-income families in industrialized and developing countries equal.

- **Volatility:** Agriculture continues to be one of the most susceptible industries when it is about natural disasters. Climate variations have instigated additionally often weather events, which can ruin a whole season's worth of harvest. Increasing temperatures are additionally resulting in the widespread spreading of crop pest. Additionally, policies of the government can influence the production of food and its prices.

For example, in the United States ethanol mandates diverted corn fields used for food production to fuel production, and resulted in price hikes from $2 or $3 to $7 a bushel. Such forces, which determine the direction of price volatility, are here to stay.

From the perspective of the DRIVE system, people can see the ways in which traditional agriculture solely will be unsustainable as a dependable and inexpensive source of food production (Figure 8.5).

Figure 8.5. A scene from inside a vertical farm.

Source: https://upload.wikimedia.org/wikipedia/commons/7/7e/VertiCrop.jpg

8.6.2. Vertical Farming Born Out of Challenges

A solution to these food supply issues is coming up from high tech structures to the dining tables of people. A term, Vertical farming coined by Dickson Despommier, is the activity of creating food in vertically-stacked layers. These "farms" utilize encircled structures like warehouses and shipping containers to offer a controlled atmosphere to develop crops in a hydroponic or aeroponic system.

Electronic sensors make sure that crops obtain the right amount of LED light, nutrients, and heat. The advantages comprise freedom from arable land, year-round growing capacities, fewer water consumption, and better-quality crop certainty.

For instance, in New Jersey, there is 70,000 square foot vertical farm known as AeroFarms in a renovated steel plant. It claims that it consumes 95% less water and 390 times extra productivity than a commercial field farm with the same square footage.

The company Growtainer trades easy-to-operate 20- or 40-foot shipping containers set up as insulated hydroponic farms. The main objective here is to assist communities to develop leafy vegetables in the same place of their consumption, such as food banks, schools, restaurants, and military bases.

In order to fulfill the needs of increasing population, Vertical farms can help by providing an extra method to produce food which does not share similar instability and risk as traditional agriculture. Despite the fact that vertical farms need less water and arable land than traditional farms, they are not carbon neutral. Their climate footprint is contingent deeply on the source from which they draw their electricity to power lighting and control the indoor environment.

As renewable energy sources develop to be adopted more extensively, the carbon cost of vertical farming will remain to be lessening. From the viewpoint of the market, it may not reduce prices, but on a communal level, the expectation is that vertical farming can benefit in dealing gaps in complete food demand where traditional agriculture flops.

8.6.3. Tasty Prospects But Not One-Size-Fits-All

For business farmers keen on venturing into vertical farming and social entrepreneurs who see the potential for utilizing vertical farming to address neighborhood food and craving issues, there are approaches to limit the costly expectation to learn, adapt, and enhance their odds of achievement:

- **Change the impression of the farming work:** Customary farming has been described as work concentrated and remote to a cutting-edge and urbanized way of life. In a few spots, farm work is related to destitution and disengagement, however, in the vertical homestead, agriculturists must be information examiners, bio-researchers, and framework chiefs notwithstanding working with crops.

Should urban homesteads keep on scaling, this could result in the relocation of existing low-talented work. Such a move is run of the mill of any significant industry change—financial specialists call this the bounce back impact. Understanding this change in farming gives experts who are either entering or as of now in the vertical farming industry with use when conveying the need to grasp vertical farming with various partners.

- **Teach purchasers:** Vertical farming is not Frankenstein food, yet should be with no endeavors to instruct the general population. Organizations can utilize limited time crusades to illuminate the estimation of non-field farming crops and teach shoppers about the wholesome and ecological advantages of vertical farming. Food-tasting occasions can likewise give customers the chance to test hydroponic and aeroponic deliver and judge the desire for themselves.
- **Bolster neighborhood food economies:** Governments and industry gatherings can be profitable partners who see nearby food creation as a financial improvement. In Canada, for instance, the Ontario Ministry of Agriculture, Food, and Rural Affairs (OMAFRA) effectively bolster a provincial food procedure, with venture bolster extending from limited time crusades to the formation of ranchers' markets and financing for horticultural organizations to purchase new gear.

With OMAFRA's help, the segment included \$1.3 billion in GDP and made in excess of 34,000 occupations somewhere in the range of 2013 and 2015. Such government bolster is an indication that nearby food developments are a believable wellspring of financial advancement without any indications of subsiding.

- **Encourage continued investment:** Financial specialists are willing to helping vertical farming scale. While some real interests in vertical farming areas of now happening—Silicon Valley startup Plenty as of late got \$200 million to help its worldwide extension—others may need to strategize more, especially since some vertical homestead new businesses have bombed in that same time allotment.

AeroFarms, for example, anchored value financing of $95.8 million by situating itself not as a nontraditional ranch yet rather as "An urban agriculture and cleantech company." Other patterns that are drawing in venture incorporate utilizing vertical farming innovation to develop supplement particular yields like Fujitsu's low-potassium lettuce.

- **Revamping the future of agriculture:** In spite of the fact that vertical homesteads can never be relied upon to supplant conventional ranches, it is likely that they should supplement each other in the event that we are to meet the food requests of tomorrow. It is monetarily sensible, naturally well disposed of, educated, and above all, wellbeing delicate. Vertical farming is not an imaginary thing; it is going on now.

8.7. CASE STUDY: A LOOK AT TEN ENTREPRENEURS WHO ARE CHANGING THE PERSPECTIVE IN AGRICULTURE

Today's young food and agricultural entrepreneurs are developing innovative ways to revolutionize the entire food chain. They are at the forefront of reducing food loss and waste, increasing crop yields, improving market access, developing novel technologies, and increasing urban and sustainable farming practices across the globe. They come from diverse backgrounds, with unique insights, perspectives, and approaches to tackling global food system challenges.

Food Tank is highlighting 10 young food and farming entrepreneurs who have transformed their love for food, farming, and sustainability into tools to build a stronger and more equitable food system.

8.7.1. Jamila Abass, Kenya

Jamila Abass is CEO and co-founder of M-Farm, a virtual co-operative for subsistence farmers in Kenya to help them reach buyers and access vital market information via SMS. Through M-Farm, farmers in the same areas can share their experiences and advice, pose questions to industry experts, and connect with each other to combine crops and find larger buyers. Abass was elected for the 2013 Ashoka Fellowship, 2015 Aspen New Voices Fellow, and as Quartz Africa Innovator (Figure 8.6).

Figure 8.6. Jamila Abass.

Source: https://c1.staticflickr.com/9/8463/8100271036_b75c3804c7_b.jpg

8.7.2. Onyeka Akumah, Nigeria

Onyeka Akumah is the founder of FarmCrowdy, Nigeria's first digital agriculture platform that connects small-scale farmers with sponsors who invest in the full farm cycle. Sponsors' funds are used to secure farmland and insure farmers and their produce from seed through to harvest. Once the crops are sold at market, profit is split between the farmer, farm sponsors, and FarmCrowdy. Sponsors can receive updates throughout the farming process via text, pictures, and video. The company has more than 1,000 sponsors across Nigeria, the United States, and the United Kingdom that support more than 2,000 small-scale farmers across eight states in Nigeria.

8.7.3. Alli Cecchini, United States

Fourth-generation farmer Alli Cecchini is the founder and Executive Director of First Generation Farmers (FGF). Providing after-school programs, summer camps, farm tours, and Willing Workers on Organic Farms (WWOOF) placements, FGF draws hundreds of people to their farm in California each year.

FGF also run their Urban Edge Sustainable Farming Program, an immersive, full-time residential program for aspiring organic specialty crop farmers. The program provides training in foundational production skills, business knowledge, and personal development to establish the next generation of resilient urban farmers and farm enterprises.

8.7.4. Alfredo Costilla Reyes, Mexico

Electrical and computer engineering graduate student, Alfredo Costilla Reyes, is the founder of BitGrange, a hydroponic system that can be sustained using only an LED light. Using novel software, BitGrange evaluates environmental variables, such as temperature and light, in real time and notifies growers through BitGrange's iPhone app to take necessary actions, such as adding more water or plant food.

Through its four-step process of "Plant-Connect-Sync-Play," BitGrange aims to gamify agriculture for its users, encouraging, and engaging youth to participate in agriculture. In recognition of his development, Costilla Reyes was awarded the Mexico National Youth Award, the highest award presented by the Government of Mexico to the country's youth.

8.7.5. Luke Craven, United Kingdom

Luke Craven is a co-founder of BiJimini, an award-winning start-up that farms crickets and produces a high-protein cricket blended flour, PowerFlour. PowerFlour combines organic white and wholemeal flours sourced from local millers with finely milled cricket powder. The flour is rich in vitamins and minerals and is an alternative to traditional forms of animal protein.

Cricket farming emits fewer greenhouse gases, requires less space, and less feed than conventional livestock. BiJimini aims to promote its flour product and pure cricket powder to consumers searching for a sustainable source of protein.

8.7.6. Matt Fountain, Scotland

Matt Fountain is the founder of Freedom Bakery, a social enterprise and artisan bakery aimed at reducing the reoffending rate in Scotland through providing on-the-job baking training for prisoners and ex-offenders. Through an apprenticeship program, graduates attain an industry-recognized qualification to help increase employability and decrease the risk of reoffending.

For some of its products, the bakery sources ancient grains indigenous to the Scottish Highlands and utilizes Scottish artisan baking techniques. Freedom Bakery is the first social enterprise to establish itself within a prison in Scotland and is now expanding outside of prison to open a second bakery, Freedom 2.

8.7.7. Josh Gilbert, Australia

A farmer first and foremost, Josh Gilbert is also an advocate and change maker in Australian Indigenous, agricultural, and environmental fields. Gilbert creates a range of campaigning and educational materials, including Climate Reality Project video 'Australia's Young Green Farmers,' podcast Tractor Talks, and upcoming cookbook collaboration with the World Wildlife Fund showcasing the stories of 50 Australian farmers.

As the Chair of the New South Wales Young Farmers' Council, he moved one of the first international agricultural climate change motions in 2016. For his advocacy, he has been recognized as the Australian Geographic Young Conservationist of the Year and the Foundation for Young Australians' Local Legend Award.

8.7.8. Dean Hale, Canada

Dean Hale is the founder of Growing Green Hamilton, which aims to advance urban agriculture by helping people grow their own produce through Square Foot Gardening (SFG). SFG is a method for people with limited space to plan and create a small, but intensively planted vegetable garden.

Hale helps members in the local community to cultivate square foot gardens, providing a nutrient-rich soil mix that is weed-free, requires no fertilizers, and can grow up to five times the amount of produce than conventional methods. Hale has connected with local community groups and churches to set up gardens to either sell the produce to raise funds or donate it directly to food banks.

8.7.9. Kate Hofman, United Kingdom

Kate Hofman is co-founder and CEO of GrowUp Urban Farms, a community aquaponic urban farm that produces fish, salads, and herbs for a local London market. Their GrowUp Box is capable of producing more than 950 pounds of sustainable salads and herbs and 330 pounds of fish each year. Hofman plans to further develop and build GrowUp Boxes across London to provide local communities with the opportunity to learn about sustainable agriculture and provide local employment for young people who may have dropped out of formal education or are unable to find employment. The London Sustainable Development Commission selected her as a "London Leader" in 2015.

8.7.10. Justin Kamine, United States

Co-Founder and Partner of KDC Ag, Justin Kamine is growing technologies that actively shift the agriculture world toward a more Circular Agriculture Society where there is zero waste. KDC Ag has developed a technology with California Safe Soil that converts more than 30 tons of fresh food waste into fertilizer and animal feed in just three hours. Kamine serves on the board of California Safe Soil, is an investor in Aerofarms, and is an advisor to Food Future. He was recently the winner of the Clean Tech Equity Award, presented by the Prince of Monaco, for the most environmentally progressive technology.

REFERENCES

1. Abdullah, A., & Sulaiman, N., (2013). Factors that influence the interest of youths in agricultural entrepreneurship. *International Journal of Business and Social Science*, [online], *4*(3). Available at: http://ijbssnet.com/journals/Vol_4_No_3_March_2013/31.pdf [Accessed 18 Aug. 2018].
2. Agecon.unl.edu. (2014). *Future-Focused Entrepreneurship | Agricultural Economics*. [online] Available at: https://agecon.unl.edu/cornhusker-economics/2014/future-focused-entrepreneurship [Accessed 18 Aug. 2018].
3. Bharti, A., (2015). *The business of Agriculture*. [online] Businessofagriculture.com. Available at: http://businessofagriculture.com/innovation/innovation09102015.html [Accessed 18 Aug. 2018].
4. Esposito, M., Tse, T., Soufani, K., & Xiong, L., (2017). *Feeding the Future of Agriculture with Vertical Farming (SSIR)*. [online] Ssir.org. Available at: https://ssir.org/articles/entry/feeding_the_future_of_agriculture_with_vertical_farming [Accessed 18 Aug. 2018].
5. *Food & Business Knowledge Platform*, (2016). The role of agri-entrepreneurship and farming innovations for youth in agriculture engagement – Food & Business Knowledge Platform. [online] Available at: http://knowledge4food.net/role-of-agri-entrepreneurship-and-farming-innovations-for-youth-in-agriculture/ [Accessed 18 Aug. 2018].
6. Food Tank, (2018). *16 Young Entrepreneurs Building a Better Food System*. [online] Available at: https://foodtank.com/news/2018/01/young-food-entrepreneurs/ [Accessed 18 Aug. 2018].
7. Hammerich, T., (2018). *Investing in Agriculture (For the Rest of Us) – Future of Agriculture*. [online] Future of Agriculture. Available at: https://futureofag.com/investing-in-agriculture-for-the-rest-of-us-b62a12727e8e [Accessed 18 Aug. 2018].

INDEX

R

S

T

U

V

W

Y